Nicolas · The Mid-Oceanic Ridges

Adolphe Nicolas

The Mid-Oceanic Ridges

Mountains Below Sea Level

With 92 Figures

Prof. Adolphe Nicolas
Université des Sciences et Techniques
du Languedoc de Montpellier
Laboratoire de Tectonophysique
Place Eugène Bataillon
F-34095 Montpellier Cedex 05

Translator:
Dr. Thomas Reimer
Via Lavizari 2a/BSM
CH-6900 Lugano

Title of the French Edition:
Adolphe Nicolas: Les montagnes sous la mer

Avenue de Concyr – B.P. 6009
F-45060 Orléans Cedex 2

ISBN 3-540-57380-1 Springer-Verlag Berlin Heidelberg New York
ISBN 0-387-57380-1 Springer-Verlag New York Berlin Heidelberg

CIP data applied for

Typesetting: Mitterweger Werksatz GmbH, Plankstadt
SPIN 10127422 32/3130 – 5 4 3 2 1 0 – Printed on acid-free paper

. . . to the people of Oman

Foreword

Over the past 20 years, geologists have come to realize that the real object of their studies is the Earth, and that their favorite subjects, whether basalts, earthquakes, or the Tibetan plateau (for example) are only the means to understanding the Earth itself as a complete entity. Geology has thus acquired a global perspective.

The study of any particular regional problem is only of general interest in providing good foundations to investigate general phenomena. But this same particular problem can only be completely understood if its global context is taken into account. The geology of the Andes is only one of many examples of subcontinental subduction and, as such, its study contributes to our understanding of the mechanism of subduction. However, at the same time, understanding the geology of the Andes is today no longer possible without relating it to the geological history of the eastern Pacific, reconstructing the relative movements of the East Pacific Rise and the Nazca Ridge against South America and, in a wider sense, in the context of the history of the Pacific Ocean as a whole.

Geology today is based on an infinite variety of natural examples which in the general frame of plate tectonics deals with geometric objects, historic situations, or the various methods employed. The object "Earth" includes not only subjects like seismology, structural geology, and geochemistry, but also studies of the crust, mantle and core, oceans and continents, and mountain ranges and basins.

Thus, the work of the geologist has changed. In the past, he sought through methods, frequently more artistic than scientific, to classify, regroup, or combine vast and unconnected masses of data. Today, he can only interpret stated facts and further the state of knowledge by contributing his own speciality. As in medicine, the generalist who, yesterday was holding all knowledge, must make way for the specialist. The present book by Adolphe Nicolas is a brilliant example of this approach.

The author is a great specialist, perhaps even the best in his field. His home ground is the study of natural deformations in a very special and important rock type called peridotite. Although rare on the surface of the Earth, this rock accounts for nearly one quarter of the volume of our planet, and beneath our feet at a depth between 30 and 600 km, the terrestrial mantle is made up of peridotite. Knowing the laws by which peridotite is deformed means understanding how the mantle may flow, change, and transport the famous plates and, with them, the continents.

Whereas on the scale of days or years, a peridotite will be a hard, resistant rock difficult to break with a geological hammer, on a geological time scale it reacts as a plastic material deformable like some type of liquid, permitting great movement to take place in the mantle. These peridotites cause continents to drift and oceans to open, create volcanoes, trigger earthquakes, and concentrate useful minerals in large deposits. Although these movements with only a few centimeters per year are indeed slow, they are inexorable and spectacular, because the course of geological history accelerates or slows them down against a background of billions of years.

Between the investigation of a few olivine crystals under the microscope or some rocks extending over a few hundred meters and the functioning of all terrestrial movement there appears to be an immense gap. However, here, as in other instances, it is the infinitely small that permits us to explain the infinitely large. Thus, detailed observations of rocks in the field and under the microscope will enable us to understand the major geological structures of the Earth, their textures, their origin and even their history. It is certainly not unimportant which rock to choose, and Adolphe Nicolas has not selected the peridotites by accident; he progresses from the particular to the general.

The central part of this book deals with oceanic ridges, a network of submarine relief ranging over nearly 4000 m in elevation and extending more than 50000 km as an anatomosing ribbon over the ocean floors. It represents a submarine mountain range invisible from the surface. Adolphe Nicolas decided to uncover its secrets by remaining on dry land and applying the methods of a terrestrial geologist.

This approach is rendered possible because of the existence of massifs of a particular rock assemblage, the ophiolites, which are, in fact, slivers of oceanic crust transported onto the continents by mysterious tectonic phenomena referred to by the general term obduction. Ophiolites are the main theme of this book, but not to the extent of blurring its real message. This relates to the creation of oceanic ridges and thus to plate tectonics, but also to the mechanisms controlling the latter, the birth of magmas and the genesis of the major structures; in short, to the entire animated life of our planet.

This book shows, however, that having contributed precise knowledge on a certain subject does not exclude the possibility of embracing vast knowledge, a statement that holds true also in the inverse direction. The mere fact of having contributed to a discovery brings with it a way of reasoning which quickly gives access to many related subjects. Those who do not contribute to the advance of science can possess no understanding of the basis of our knowledge and find it difficult to distinguish between solid fact and disputable or provisional ideas. Science is not built on certainties but relies on uncertain hypotheses assembled into a coherent framework. Although coherent but often, on further research, shown to be provisional, this framework can later be replaced by a better, more general and more global, explanation.

This spirit of scientific research pervades the entire book of Adolphe Nicolas. The reader recognizes the continuous desire to place detail, essential observation, and elaborate modelling into a more general context, in order to understand the functioning of the system "Earth".

However, this constant desire to pass from the particular to the general, to think in global geological terms, by no means affects the originality of this book. Originality in presentation is combined throughout with the vivid expression of the author's personal viewpoints. Although there is a constant change from observed facts to theoretical construction, the book runs smoothly, illustrating the very essence of scientific progress. This wealth of ideas and hypotheses shows that modern geology is a living science, allowing the exchange of ideas and proposals.

However, throughout the book the reader recognizes the extraordinary pleasure which motivates the author in his geological work.

Having enjoyed the privilege of observing Adolphe Nicolas carrying out his fieldwork with his characteristic disciplined enthusiasm, I can recognize this spirit clearly in his book. It is this enthusiasm, this desire for knowledge, which he has passed on to the young scientists of his group, leading to the establishment of his school.

May this book inspire many young people to practice and enjoy the profession of geology.

Claude J. Allègre

Preface

Plate tectonics, rising from the oceans some 25 years ago, has revealed to us that the driving forces of activity for our Earth have been mostly hidden well below the ocean floors. Did you know that the most important mountain chain of this planet is located on the sea floor, over which it winds its way for some 75000 km?; and that the oceanic ridges making up this chain furnish at any point in time eight times the volume of basaltic lavas produced by all volcanoes, thereby forming the oceanic crust and creating over some 250 Ma years an area equivalent to the entire surface of the Earth?

Did you know that under these ridges there are caverns filled with over 1000 °C hot magma, immense cathedrals of fires, whose top reaches up to 2 km below the ocean floor?

Did you know that the same ridges and their flanks are cooled by a circulation of sea water which suffices to filter the entire ocean waters within a few million years?

And did you know that when these ridges were "gripped by fever", as during the Cretaceous about 100 Ma ago, their spine swelled to displace the oceans, which as a result of the ensuing rise in sea level invaded the borders of the continents? Geologists call this a marine transgression.

Not only scientists of numerous nations, but also the citizens, are engaged in the formidable task of making this silent world talk. Such research is expensive, involving high sea ships, manned or unmanned submersibles and sophisticated instruments capable of sounding great depths. One day expenses for such a submersible costs as much as a luxury car! Is it not legitimate for a scientific witness, albeit frequently only an indirect one, of these journeys over the oceans and of the spectacular results of this research, to allow those to partake in the discoveries who consciously or unconsciously contributed to this cause.

Why, then, should a geologist put pen to paper who, although from a family with numerous seafaring ancestors, has travelled more over land than across the oceans? The answer to this lies in a single word: ophiolites. As we shall come back to this term throughout the book, it will suffice here to say that these formations which now crop out in the interior or along the margins of the continents originating on the floor of the oceans.

While walking across these dull and bleak landscapes we may study dry-footed the composition and structures of these ocean floors. This spectacularly successful approach complements the study of the ocean floor by the genuinely marine means referred to above.

It was one of the moments of chance in my scientific career that, after having published a thesis on ophiolites cropping out in the Alps, I should become interested in the mechanisms of rock deformation. Jean-Paul Poirier, a physicist, and myself discovered that, under the influence of heat, rocks may be deformed like metals and that consequently it would be permissible to apply the concepts of metallurgy, an old and successful science, to terrestrial deformation.

Connection with the ophiolites was not so difficult, as we possessed our own equivalent of the Rosetta Stone of the Egyptologists, namely peridotite, a rock from the Earth's mantle which constitutes the floor of the oceanic crust and forms an important portion of the ophiolites. When our knowledge of the "metallurgy of rocks" had, in my opinion, advanced sufficiently (in 1975), my group and myself directed our attention to applying these new methods with the aim of understanding the functioning of oceanic ridges. Our approach entailed mainly field studies in some 20 ophiolite massifs around the world. In these rocks we were looking for traces, albeit frozen ones, of the transport of very hot mantle below the submarine cover of the oceanic mountain ranges, where in a maelstrom of heat, oceanic crust is formed. It is the birth of this crust that I would like to talk about.

However, a word of caution is necessary here. A scientific treatise like the one presented here usually describes the state of knowledge, i. e. results which have been submitted to the proof of time, and which represent a certain consensus within the scientific community. In contrast, a number of the results given here are new and have not yet been fully elucidated, and may even be disputable. This applies especially to those results gleaned from 15 years of fieldwork in the most beautiful ophiolite in the world, the Oman ophiolite along the eastern edge of the Arabian Peninsula. These data enable us to propose a model of the functioning of ridges which is largely compatible with the most recent oceanographic data.

I have attempted to write an easily accessible book. In my student days my image of science was characterized by abstract concepts and a dry way of expression in the physical sciences and by a host of incomprehensible words in the so-called natural sciences. In the latter, things have come a long way since then. Less and less are catalogues, classifications, and profuse scriptions confounded with real science. This demystification quite naturally also led to a clearer language and to the development of physical reasoning and its mathematical expression. During our travels we shall try to circumnavigate these cliffs of abstraction and language. We can most easily avoid the abstraction of the physicists. As far as the shortcomings of language are concerned, we shall try to limit descriptions for specialists, and include at the end of the book a glossary of technical terms. The text itself contains two levels of understanding. In addition to the normal text, a second level is marked by a recessed margin and a vertical bar. It describes developments, the reading of which is not indispensible for general comprehension. Its contents are repeated in simpler terms in the first lines of the following paragraph. For those who would like to delve deeper into the subject, there is a third level to this volume to be found in my book *Structures of Ophiolites and Dynamics of Oceanic Lithosphere* published by Kluwer in 1989.

There are by now a number of excellent treatises which have made the concept of plate tectonics popular. I am thinking here in particular of *L'écume de la Terre* by Claude Allègre, now translated into English as *The Behaviour of the Earth*, which recalls the tremendous adventures of the past 20 years and calls attention to the ramifications of scientific discoveries and consequently of their ties to the history of research. It is thus permissible for me to limit the description of the more general developments to those which I feel are indispensible for fully understanding the subject on hand, viz. the genesis of the ocean floors.

Acknowledgements: A number of the results presented in this book belong to the saga of peridotites and ophiolites, which is over 20 years old, and the heroes of which are my colleagues and students from a small group at the University and at the CNRS. The skeleton of this group is made up of Françoise Boudier, Jean-Luc Bouchez, Yves Gueguen and David Mainprice, who contributed their charisma and competence. Some 20 young graduates gave their ardour and devotion. To this group I owe most. Emanuel Ball deserves special mention because of his active participation in conceiving and drawing the illustrations of this book; Solange Fournier brought it into shape, and several underwater photos were kindly supplied by Jean-Marie Auzende. The number of scientific colleagues and long-suffering friends who reread and corrected these pages is large: Jean-Paul Poirier, Françoise Boudier, Michel Prévot, Maurice Mattauer, Joël

Lancelot on one side, and especially the members of my family, Odile, Valentine, Clarisse, and Yves as well as Jacqueline Goyallon on the other.

My thanks go to all of them.

Contents

1 Young and Living Oceans

Plate tectonics, a child of the oceans, may be demonstrated as a reality with the aid of the record of the opening of the oceans as presented by magnetic anomalies. Created along a ridge, the oceanic floor increases in age as it slides towards the subduction zones where it is finally swallowed. The fire of the ridge fades rapidly away during this journey and a cold skin, the lithosphere, forms with increasing thickness above a still hot and mobile mantle, the asthenosphere. Thickened and of greater weight, lithosphere older than 150 Ma sinks spontaneously along subduction zones. In contrast to this, young, thin and light lithosphere tends to resist subduction, and sometimes quite victoriously so. This is when ophiolites form.

The concept of very ancient and immovable oceans being as old as the Earth itself has been swept away since the early 1960s by the rise of plate tectonics. The oceans are geologically young and subjected to continuous rejuvenation. Plate tectonics proved to be a veritable fountain of youth for the earth sciences, just as the impact of the discovery of DNA in biology 10 years earlier, or that of relativity on physics at the start of the century.

Most of the new concepts which have firmly established plate tectonics were furnished by the systematic study of the ocean floors during the 1950s, mainly under the auspices of the US Navy. To Mary Tharp and Bruce Heezen belongs the merit of compiling the first systematic maps of submarine relief. These submarine topographic maps are based on soundings of the depth of the ocean floor below the keel of an oceanographic vessel. On these by now well-known maps, the ocean floor appears to become deeper on either side of a median spine, the oceanic ridge, which itself is dissected by cross cutting fracture zones.

The Opening of the Oceans – Proof of Global Tectonics

The progressive opening of the Atlantic Ocean between Europe and Africa to the east and the Americas to the west had been proposed by the Germany meteorologist Alfred Wegener in 1912 as part of a theory of continental drift which was radically new at that time, when geological science was highly fixistic and even anthropomorphous in relation to time. This vision of continents in motion had an impact comparable to that of Galileo's idea in relation to space. It was not accepted because the state of the science and the reactions of some scientists tended to slow down audacious advances. Fiercely rejected by geophysicists of the period, the theory of mobility or drift of the continents was revived by an entirely new generation of geophysicists on the basis of magnetic anomalies.

The Zebra Patterns on the Ocean Floors

The rise of oceanic research since the Second World War, which furnished innumerable maps on the floor of the oceans (Fig. 1.1) resulted also from the desire to construct a map of the related magnetic fields. Towed by oceanographic vessels, magnetometers, which had been developed for the military purpose of detecting enemy submarines, facilitated the construction of magnetic maps. To everybody's surprise, it was found that on the ocean floors there was a banded pattern in which the local field is either above or below the mean regional field. The respective zones are referred to as positive or negative anomalies. These bands are further aligned parallel to the ridges observed on topographic maps. The ridges themselves are highlighted by a straight band of normal polarity (black in Fig. 1.2) framed on

Figure 1.1
Map of the submarine relief of the Atlantic Ocean. The strong vertical exaggeration highlights the culmination of the ridge, the axial rift and its dissection by fracture zones. Note also that the true ocean/continent boundary, the continental slope, appears like an escarpment at the exaggeration chosen. The platform above the slope belongs to the continental domain (After an original drawing by B. C. Heezen and M. Tharp). Map of sea-bottom by M. Tanguy de Rémur© Hachette – Guide Bleus ▶

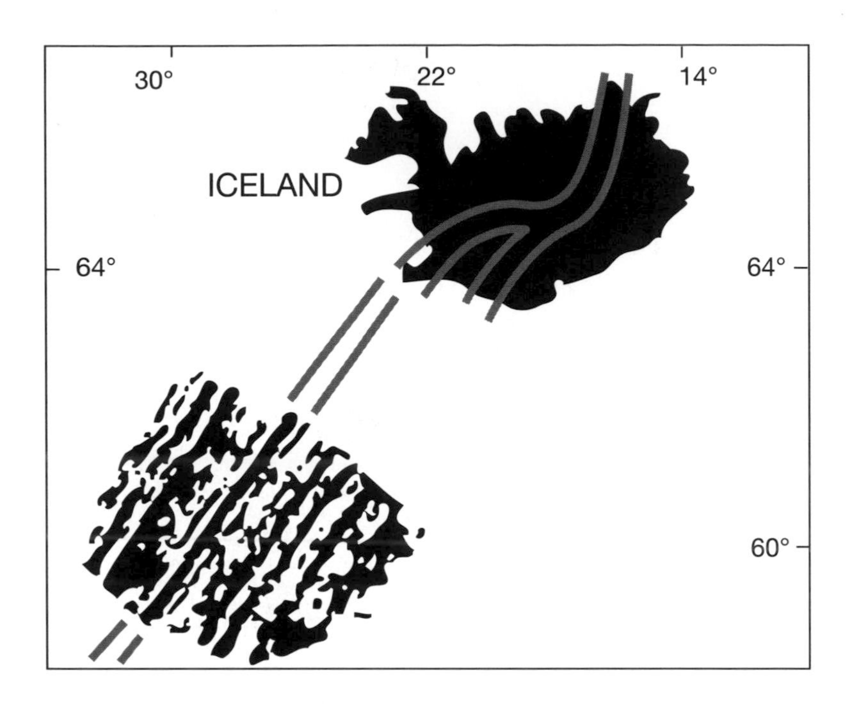

Figure 1.2
Map of the magnetic anomalies on either side of the Mid-Atlantic Ridge southwest of Iceland. The *double line* represents the ridge and its presumed extension onto Iceland

either side by bands of inverse polarity (white in the same picture). How must we understand this zebra pattern which is symmetrically developed on either side of the ridge?

It is now generally accepted that these field anomalies are caused by the exceptionally strong magnetization of the oceanic crust. To understand this, we have to remember that any magnetized body creates around itself its own magnetic field. The magnetization of basalts erupted on the ridges and, covering the ocean floors consequently, locally modifies the magnetic field of the Earth, which itself is created at much greater depth in the liquid core. When the magnetization of the basalts is parallel to the recent field, the anomaly is considered as positive whereas it is negative when the magnetization is opposed to the recent field.

What causes the magnetization of these basalts? It is simply a trace of the fossil field. The ocean floor basalts become magnetized parallel to the Earth's field at the time of their cooling on the ridge. The rock behaves like a compass, the needle of which gets stuck immedi-

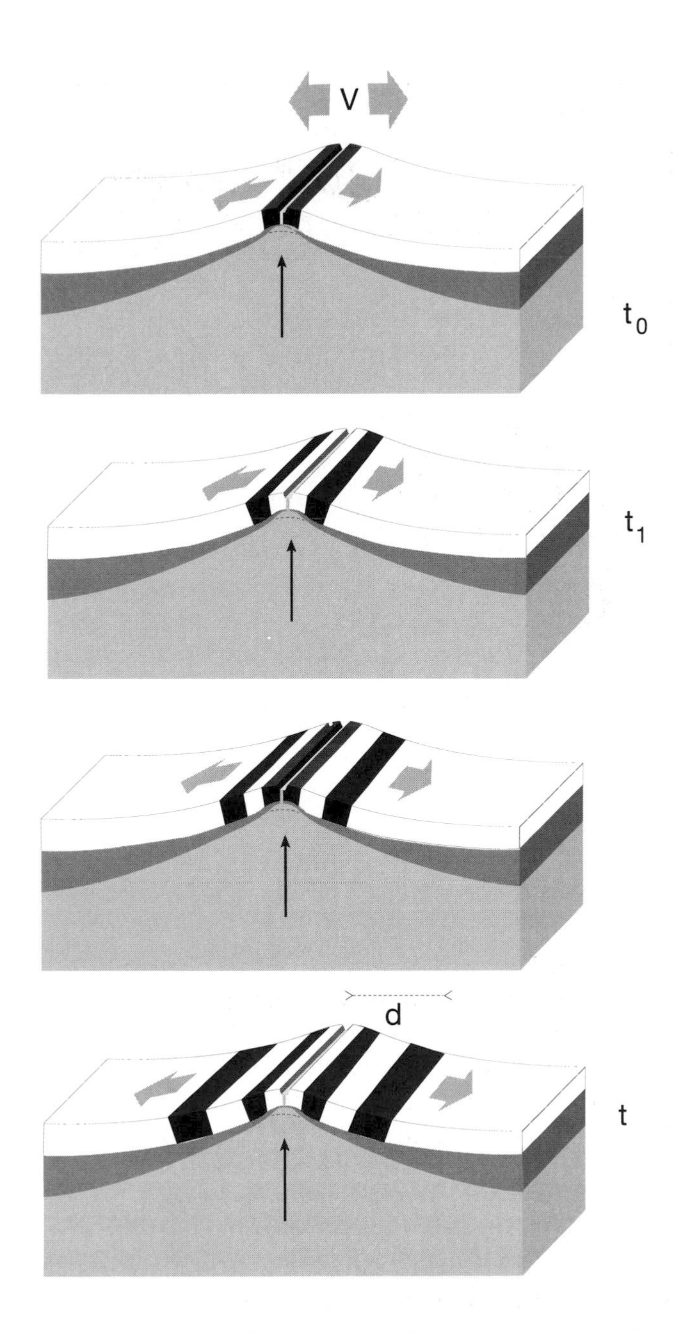

Figure 1.3
Origin of magnetic anomalies along a ridge. At the present time (t_0), a basaltic lava with normal polarity (*black band*) is extruded on the ridge until a point in time when a magnetic inversion takes place. From this time (t_1) onwards the ridge extrudes basaltic lava of inverted polarity (*white band*) which splits up and pushes to the side the older lava bands of normal polarity. The phenomenon is repeated and thereby leads to the magnetic stripe pattern of the ocean floors. The velocity of opening V is calculated by simply measuring the distance d from the ridge for a certain stripe, the age of which is known. The formula is $V = 2d\ (t\text{-}t_0)$

ately after eruption. However, the terrestrial magnetic field undergoes reversals during which its north and south poles change position at a frequency ranging from several dozen to several hundred thousand years. This still not completely understood phenomenon was first demonstrated at the start of the century and has become widely accepted since the discovery of the magnetic stripe pattern on the ocean floors. The combination of these discoveries represented the clearest demonstration of ocean floor spreading and further underlined the role played by the ridges in this process. Figure 1.3 shows that the expansion is caused by a basaltic intrusion at the ridge axis which during cooling records the orientation of the magnetic field at this particular point in time. Should the polarity be inverse, the internal compass of the lava will point towards the south. Over the range of time considered here, i. e. millions of years (Ma) or fractions thereof, the intrusion will be continuous. The young basaltic crust is then pushed to the sides to permit the rise of fresh hot basalt along the ridge. When at some time in the future a reversal of the magnetic field takes place from the present positive field, apparently instantaneously on our time scale, the new basaltic intrusions will become noticeable immediately by a band of inverse polarity parallel to the ridge axis. The normal crust will be split into two stripes on either side of the axis by the intrusion of inversely magnetized material on the ridge (Fig. 1.3). The reversals of the magnetic field thus represent spectacular markers of the expansion of the ocean floors.

Plate Kinematics – a Cartoon

Due to the fossils contained in sediments and isotopic age determinations on lavas and other rocks formed at high temperatures, we are in a position to date precisely the age of past inversions of the magnetic field. As shown in Fig. 1.3, it will suffice to establish the width of a certain stripe of known age to calculate the rates of ocean floor spreading. These range from 1 cm/a to 20 cm/a, corresponding to 10–200 km/Ma. The velocities are usually expressed by bulk rates which are double the rate of motion of one flank in relation to the ridge axis. To obtain the bulk rate, we have to keep in mind that the width of a certain stripe is half the width of the band originally formed on the axis (Fig. 1.3). These data on the opening velocities of the oceans allow us to construct so-called kinematic maps of the actual displacement of different plates in relation to each other (Fig. 3.5). One may go even further and apply this analysis to stripes which are farther

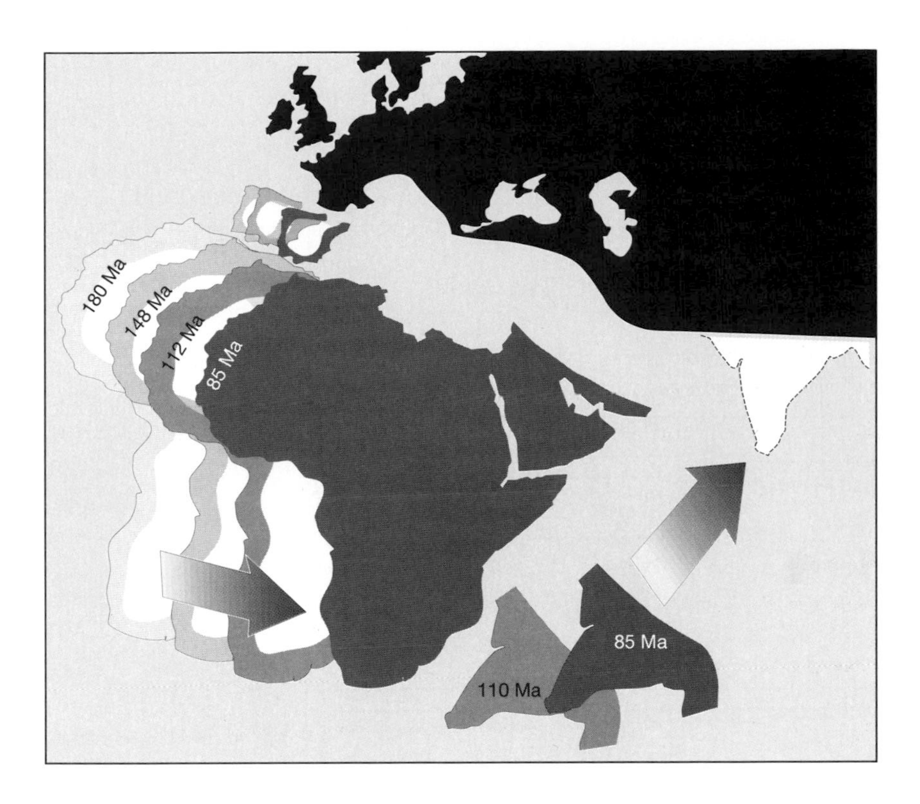

Figure 1.4
Drift of Africa and India against a Eurasia considered as fixed. The drawing is based on maps of the opening of the Atlantic (Fig. 1.5). (After Patriat et al. 1982. Bull. Soc. Géol. France, 24, 363–373)

away from a ridge and thus increasingly older, in order to reconstruct relative movements of the plates during different periods. Thus, the movement of the African plate against the European plate (Fig. 1.4), which is considered as fixed, a situation not far from true, may be deduced from maps of the progressive opening of the Atlantic (Fig. 1.5).

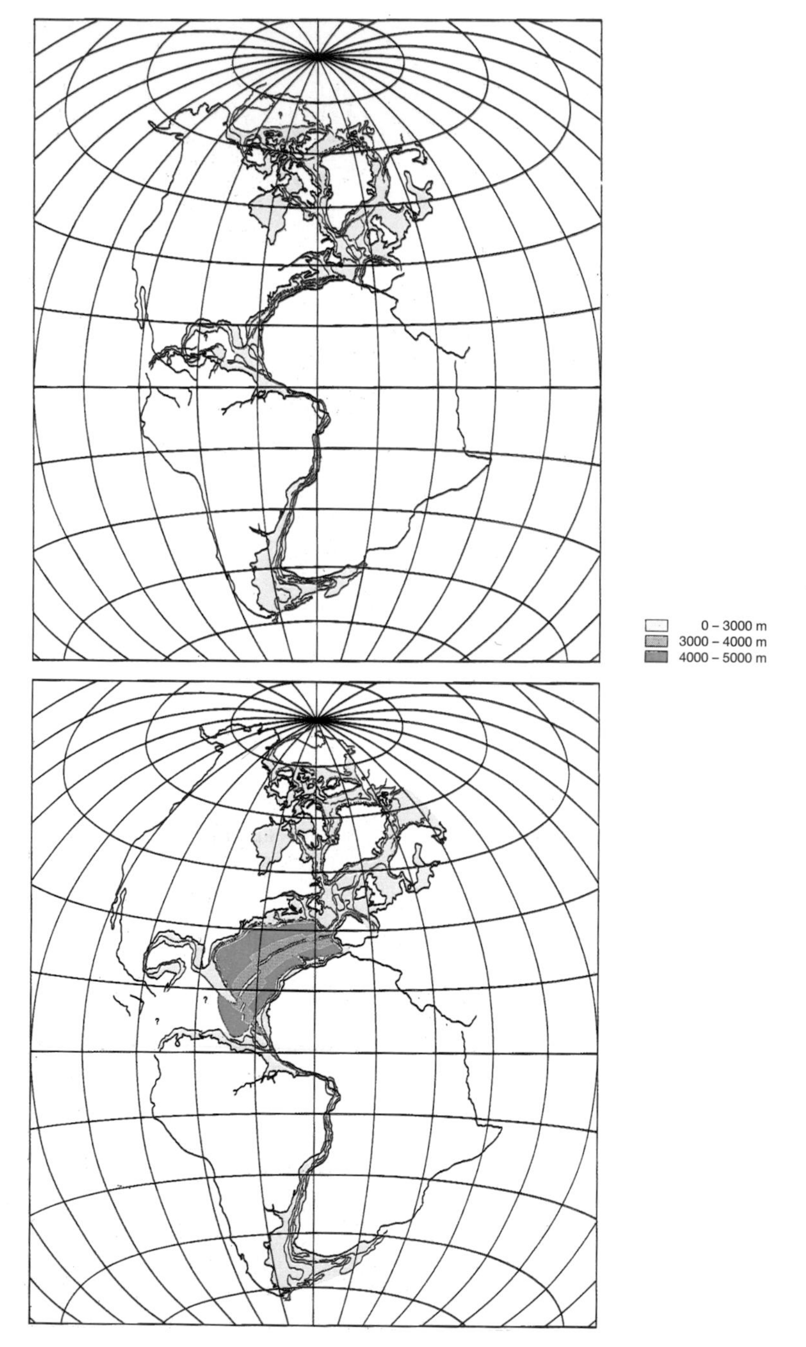

Figure 1.5

Reconstruction of the continental drift since the Jurassic (165 Ma), based on an analysis of the magnetic anomalies in the Atlantic. Assembled until 165 Ma in the super-continent Pangea (cf. Chap. 9), the continents started their drift with the

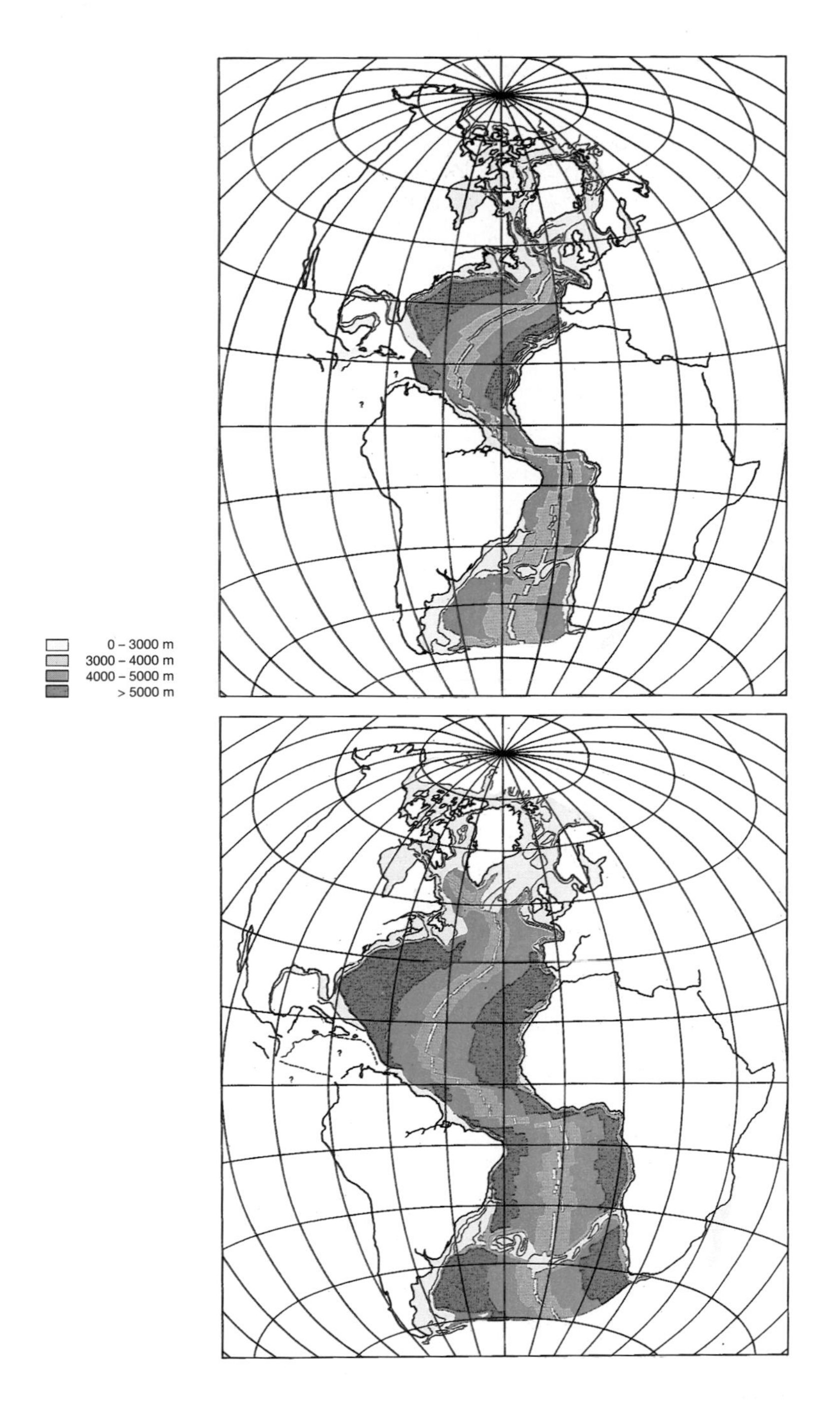

opening of the North Atlantic, followed from 125 Ma onwards by the opening of the South Atlantic (J. G. Sclater and C. Tapscott 1979. La dérive des continents, Bélin ed., 106–119)

The Sedimentary Record of the Opening of the Oceans

When dating the bands of magnetic anomalies, we note that the oldest stripes are those now closest to the continents and that the youngest are closest to the ridges. This chronologic record confirming the expansion of the oceans may also be proven by the sediments covering the ocean floors. Since 1969, the marine scientific community has had at its disposal a scientific drilling vessel within the framework of a programme known initially under the acronym DSDP (Deep Sea Drilling Project) and now as ODP (Ocean Drilling Program). The samples collected from drillholes in oceanic sediments facilitated the confirmation of the above age relationship: on the ridge itself the crust is "naked". Farther away from the ridge, it becomes covered by an increasing thickness of sediments, the bottom of which at the same time increases in age. Fossils contained in these sediments show that the oceanic crust is nowhere older than the Jurassic, or about 180 Ma.

Life and Death of the Oceans

Before spending more thoughts on the cradle of the ocean, we should think about their ultimate destiny. The oceanic crust formed along the ridge expands in a continuous regular fashion as shown by the analysis of the magnetic stripe pattern. Unless its journey over the sea floor is somehow interrupted, it will extend over more than 150 Ma. In the course of such a long life, the oceanic crust grows older until it is eventually swallowed up in a subduction zone, a journey retraced by Fig. 1.6.

The First Signs of the Age

Immediately after having formed along a ridge, the oceanic crust starts being subjected to the process of ageing. The ocean water penetrates into and alters the crust by way of a network of dissecting fissures. It becomes increasingly covered by sediments, submarine volcanos pierce its skin like warts and fracture zones cutting across the ridge carve it up to great depth (Fig. 1.1). During spreading, the depth of the ocean floor increases regularly as it moves away from the ridge (Fig. 1.7), a situation referred to as thermal subsidence. We shall stop here for a moment, as we are now coming to the thermal

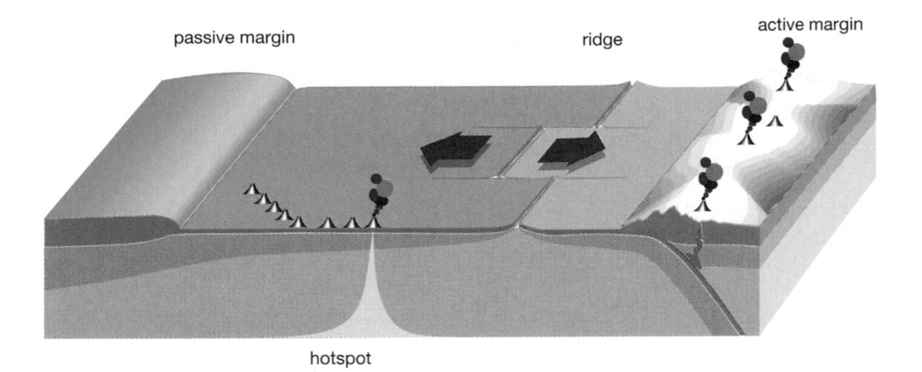

Figure 1.6
Oceanic lithosphere is extruded at its ridge of origin travels across the ocean, becoming thicker in the process, and eventually is buried along the subduction zone

structure of the ocean floor, the importance of which we shall understand later.

Thermal Subsidence

The origin of thermal subsidence has to be sought in the principle according to which an object usually expands on heating and contracts on cooling. We shall see that at the base of the skin under the ridge (-2 km) temperatures of 1000 °C and above are attained. As the formations move away from the ridge, they start to cool. Within the first few kilometres below the top of the oceanic slab, this cooling takes place very rapidly due to the circulation of sea water (Chap. 4). Below this, cooling will take place less rapidly as the sea water has more difficulty in penetration and the heat is increasingly dissipated by diffusion through a solidified environment. This type of heat transfer, thermal conduction, brings us to the phenomenon of diffusion, which obeys a simple law of $e = \sqrt{Dt}$. In this, e is the distance covered by diffusion over the time period t taking into account the diffusion coefficient D, which expresses the resistance to transfer encountered in a certain environment. As we move away from the ridge, the thickness of the cold solidified skin covering the deep, hotter, and more mobile portions increases progressively. This skin is referred to as the lithosphere (sphere of rocks) in contrast to the underlying asthenosphere (sphere of weakness Fig. 1.9). Starting from the above-mentioned law of dif-

law of diffusion, we may calculate the thickness, e, achieved by the lithosphere at the expense of the asthenosphere in an oceanic slab of age t. The calculation becomes somewhat complicated as temperatures in the lithospere grade from 0 °C at the ocean floor to 1000–1100 °C in the transition zone to the asthenosphere and because the diffusion coefficient depends on the temperature. We accept here a simplified relationship $e = 9.5 \sqrt{t}$ (km/Ma). Where we cannot precisely determine the thickness of the lithosphere, we still know the depth of the ocean, which, in turn, is tied directly to the thickness of the lithosphere as it results from the thermal contraction and related increase in density which are produced when hot asthenosphere is transformed into lithosphere by cooling. This explains why, when plotting ocean depth Z against its age $\sqrt{t}$, we obtain a straight line, at least until about 60 Ma (Fig. 1.7). Beyond 60 Ma, the floor of the oceans deepens less quickly than expected. The simplest way to explain this flattening of the sea floor is to admit that for great ages the lithosphere received more heat from below, but the reasons for this change in behaviour are still being discussed. The limiting thickness of lithosphere is $e = 9.5 \sqrt{100} \sim 100$ km.

Thus at a given point, the depth of the oceans reflects the thickness of the underlying lithosphere. This cold and solidified layer is not hot enough to become deformed easily. It is also referred to by the term "plate", overlying the hot and plastic asthenosphere (Figs. 1.7 and 1.9). With increasing age, the lithosphere becomes thicker as a result of the loss of heat to the oceans, just as a in cooling bath of molten paraffin the fluid is covered with a crust which progressively increases in thickness due to the loss of heat to the overlying air. This analogy is somewhat dangerous, as the asthenosphere is made up of mantle material which, although certainly hot and deformable, is in a solid, non-liquid state. The thickening of the lithosphere at the expense of the hotter asthenosphere leads to a contraction and thereby progressive deepening of the oceans with age. The depth of the ocean floor below sea level thus grades from -2500 m above the ridges to about -6000 m in the abyssal plains. At the same time, the thickness of the lithosphere increases from a few to about 100 km. For the same reasons, there is also lithosphere below the continents where its thickness is also 100 km, although it may be much thicker below very old crust.

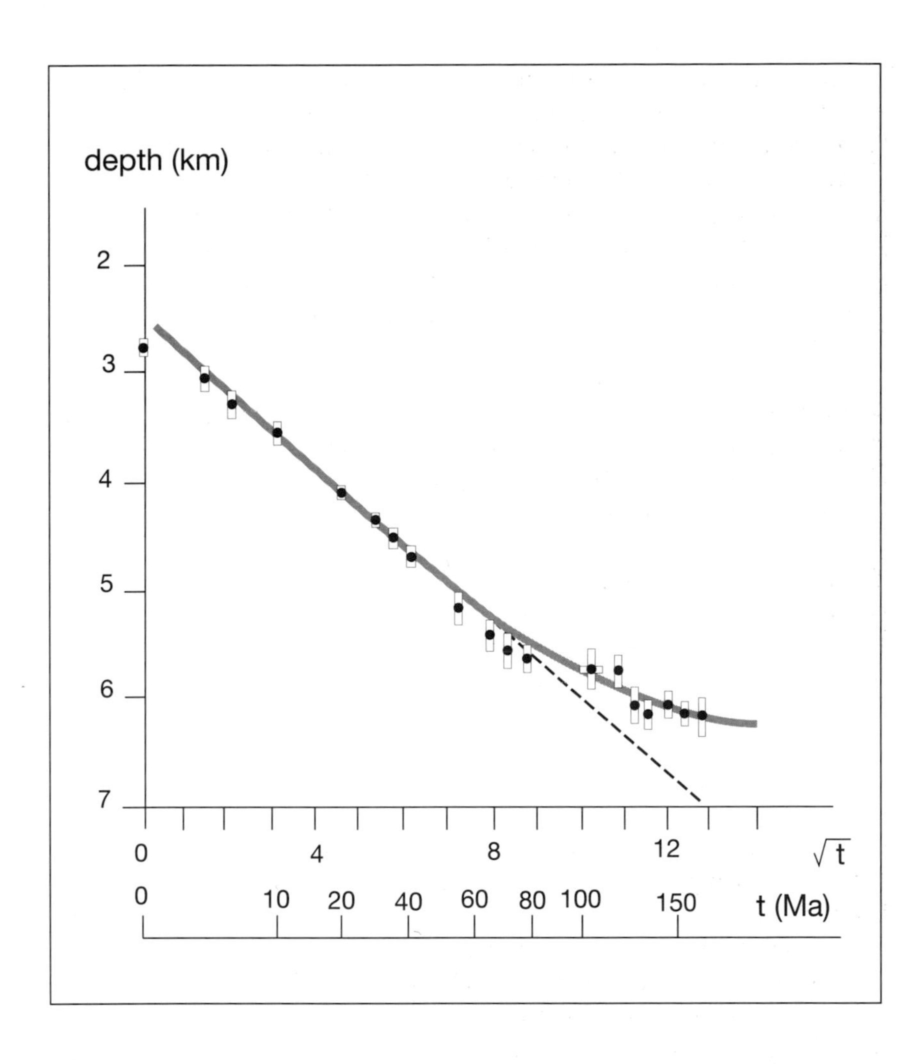

Figure 1.7.
Depth of the floor of the Pacific Ocean vs. age of the lithosphere (expressed here by the square root of the age in million years). The dashed line corresponds to the square root curve which for a value of 7.7 gives a corresponding age of t = 60 Ma. (After B. Parsons and J. G.Sclater 1977. J. Geophys. Res., 82, 803–827)

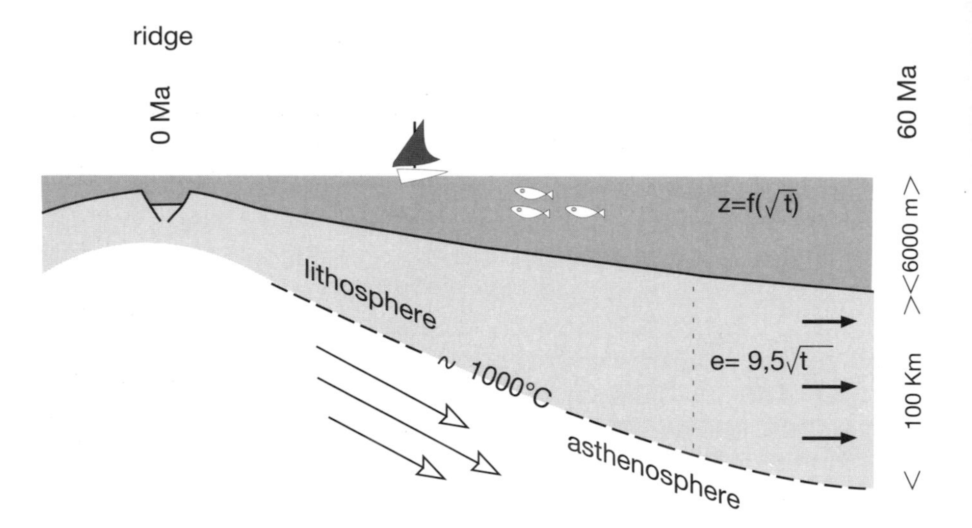

Figure 1.8
Rigid oceanic lithosphere is formed along the ridge by cooling of hot ascending plastic asthenosphere. The lithosphere becomes thicker with age (thickness e) until about 60 Ma and then stabilizes around 100 km. As a consequence, the depth of the oceans increases to stabilize eventually at about 60 Ma and a depth of about 6000 m (cf. Fig. 1.7)

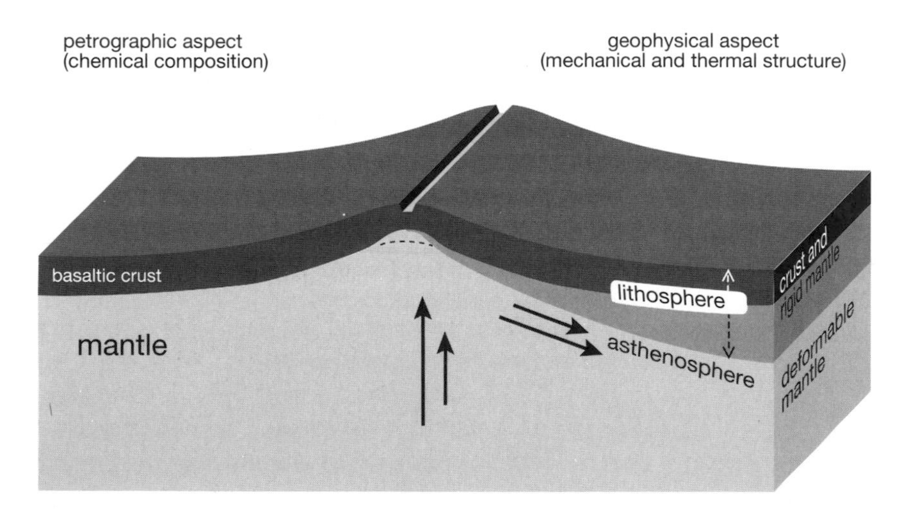

Figure 1.9
Petrographic aspect (chemical and mineralogical composition) and geophysical aspect (physical structure) of the ocean floors. Crust and mantle are chemically and mineralogically different entities. The concept of lithosphere vs. asthenosphere is based on differences in temperature and mechanical behaviour

Maturity

We have defined the lithosphere as a purely physical entity: it is the cold, rigid surface layer of the Earth. In the immediate vicinity of a ridge, the lithosphere may consist only of oceanic crust, which is basaltic in composition. However, as one moves away from the ridge for more than a few dozen kilometres, the "mantle" component grows rapidly (Fig. 1.9). At first sight, we have to admit that the only difference between mantle incorporated on cooling into the lithosphere and the underlying asthenospheric mantle lies in their respective densities. The colder lithospheric mantle layer is denser than the underlying asthenosphere. As we know that the oceanic crust is lighter than the underlying mantle, it is possible to calculate the thickness of an oceanic lithosphere with its lighter crustal and denser mantle portions which will be in exact equilibrium with the asthenosphere, i.e. it will neither rise nor sink. We can illustrate this by a weight on a scale (Fig. 1.10a). Knowing the thickness of this lithosphere in equilibrium, we may calculate its age as about 30 Ma. Oceanic lithosphere younger than 30 Ma floats on the asthenosphere whereas lithosphere older than 30 Ma should spontaneously sink. Becoming increasingly unstable with age, this equilibrium may nevertheless persist up to about 150 Ma because of the resistance exerted to this sinking by the asthenosphere which – as we should remember – is a deformable solid and not a liquid. Consequently, beyond an age in the order of 150 Ma, lithosphere will sink inexorably. Before embarking with the sinking oceanic lithosphere onto the problem of subduction, we shall extend our analysis to the continental lithosphere.

Old Continents and Young Oceans

From a systematic exploration of the oldest rocks in the old continents and from their radiometric ages, we now know that the first continents started to stabilise around 3800 Ma ago. Between the birth of the Earth 4500 Ma ago and these 3800 Ma, the convulsions suffered by our planet were so vicious that fragments of crust which floated like scum on the mantle became either rapidly reabsorbed or dislocated by the impact of enormous meteorites. The wrinkled face of the Moon, which "froze" a long time ago, bears witness to the violence of such impacts. After 3800 Ma, the continental crust never stopped growing and never became truly reabsorbed in the mantle. This is in amazing contrast to the oceanic crust, the age of which does not exceed 180 Ma. It is in the more or less efficient memory of the con-

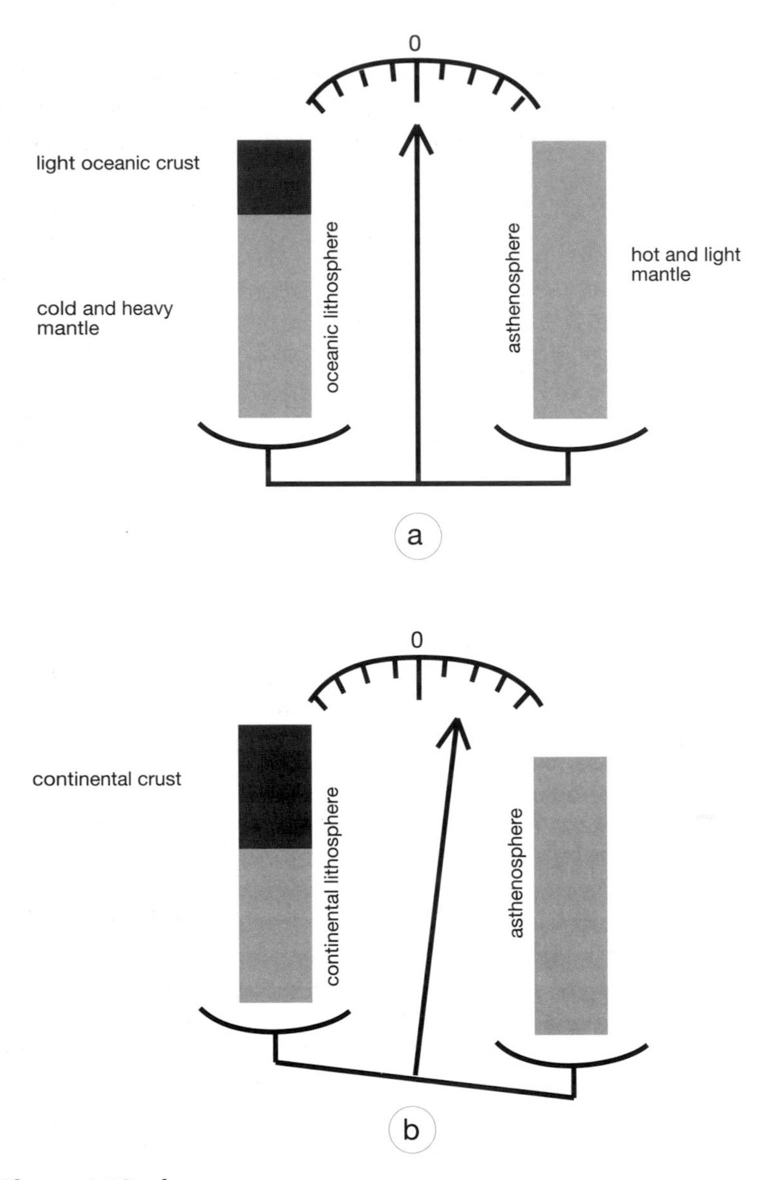

Figure 1.10 a,b
Comparison of the weight of oceanic (**a**) and continental (**b**) crust vs. a prism of asthenosphere of the same height. At the selected thickness of oceanic lithosphere (scale **a**) corresponding to a certain age, one obtains an equilibrium by which, in the oceans, the lithosphere neither sinks nor floats. Such an equilibrium is never attained by continental lithosphere (scale **b**): whatever its age, it will always attempt to float. For oceanic lithosphere the equilibrium is attained at 30 Ma. The relevant formula reads as follows: $d_l\ (e_l - e_c) + d_c e_c = d_a e_l$, with d_l = lithospheric density = 3.30; d_c = density of oceanic crust = 2.8; d_a = asthenospheric density = 3.25; e_l = thickness of lithosphere in km; e_c = crustal thickness 5 km. From the deduced value e_l=50 km and applying the law of crustal thickening with time ($e_l = 9.5 \sqrt{t}$) we obain t = 28 Ma, a correspondence which is reasonable considering the orders of magnitude involved

tinents or on the face of the still fresh Moon that we should look for "clues" to the early history of the Earth. Our knowledge of the oceans tells us more about the present life of the Earth, a subject referred to as geodynamics.

The resistance offered to subduction by continental lithosphere is explained by the fact that in contrast to oceanic lithosphere, whatever its age, it remains lighter than the asthenosphere. Actually, the continental crust is much lighter and five times as thick as the oceanic crust. It is not counterbalanced by the lithosphere, which does not thicken indefinitely with time and does not seem to exceed 200–300 km under the oldest continents. In our model (Fig. 1.10b) a column of continental lithosphere will thus always be lighter than a column of asthenosphere of the same height.

The Marine Cemetery

The presence of ophiolites, which represent slabs of oceanic lithosphere incorporated in the continents, illustrates clearly that disappearance is not the inescapable destiny of this lithosphere. However, the volume of these particular rocks accumulated on continents over several hundred million years is negligible compared to that of the oceanic lithosphere. We shall discuss in Chapter 5 how ophiolites managed to escape what otherwise remains the natural fate of the oceanic lithosphere.

The subduction zones, veritable marine cemeteries, are located at the edge of two plates which converge and thrust one over the other (Fig. 1.11). The plates concerned may both be marine, as in the subduction zone marked by the Mariana Islands and the adjoining trench in the western Pacific; or one plate may be marine and the other continental, as the subduction zone along the eastern border of the Pacific. Here, following the general rule, it is the marine Pacific plate that is buried under the continental American plate along the North American Cordillera and the Andes. Finally, both plates may be continental. We then talk about a collision rather than a subduction zone. This is the case in the collision between India and Asia which gives rise to the Himalayan mountain range.

Considering the subduction zones which surround the Pacific Ocean, those of the "East Pacific" are frequently contrasted with those of the "West Pacific" (Fig. 1.11). The former affect young or relatively young oceanic lithosphere (Fig. 1.12) whereas the latter affect a lithosphere which has already reached its age limit of 150–200 Ma. The eastern Pacific subduction zones are "forced" subductions, i. e.

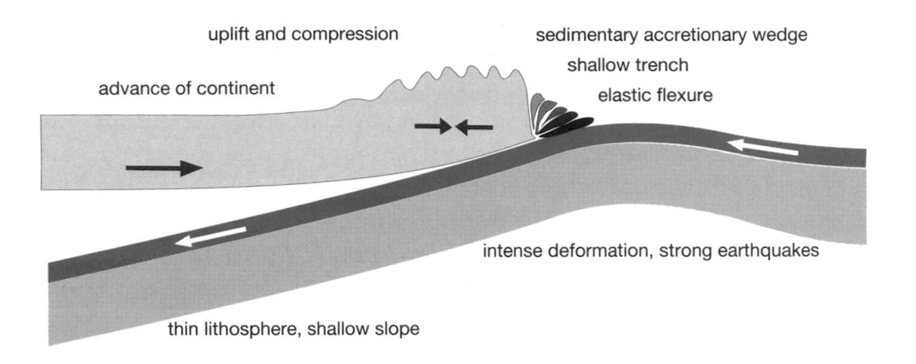

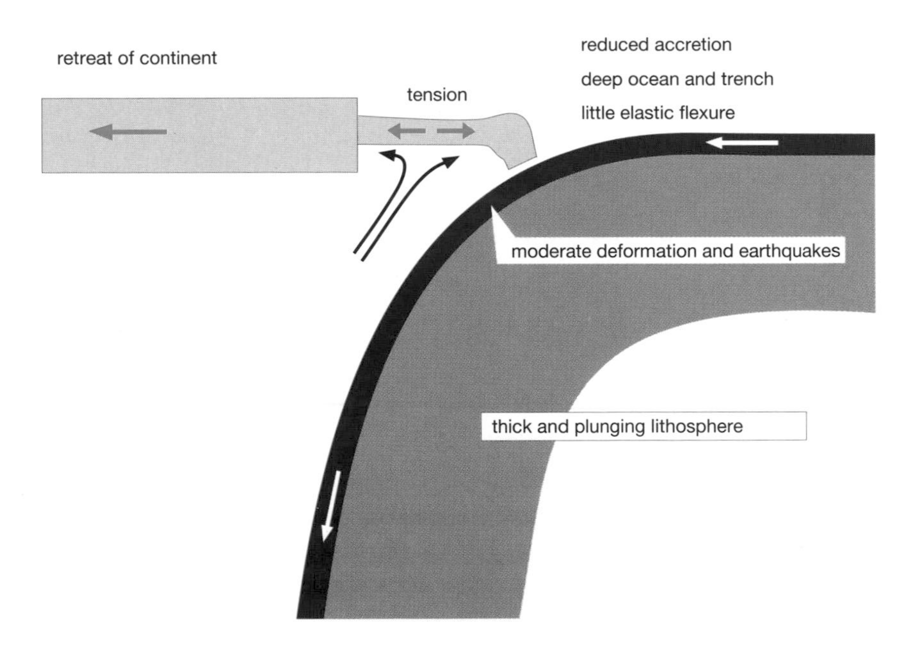

Figure 1.11
Theoretical sections through different types of subduction zones: forced subduction of young lithosphere like that of the Pacific Nazca Plate under Chile (*top*) and spontaneous subduction of old lithosphere in the Marianas (*bottom*). (After S. Uyeda and H. Kanamori 1979. J. Geophys. Res., 84, 1049–1061)

in order to enter the subduction zone, the lithosphere has to be pushed from behind whereas in the western Pacific "spontaneous" subductions, oceanic lithosphere plunges on its own. We have advanced beyond an explanation for this contrasting behaviour: young lithosphere, theoretically younger than 30 Ma is lighter than asthenosphere and thus resists sinking along a subduction zone. The very old lithosphere sinks spontaneously, as its density is higher than that of the underlying asthenosphere. As expected, spontaneous subduction results from causes intrinsic to the lithosphere itself and not from external causes. It is thus independent of the geodynamic situation and may occur even in the open ocean. This could explain the location of the western Pacific subduction zones in an oceanic environment.

After its long lifespan, oceanic lithosphere will bury itself in the grave of a subduction zone, which it dug itself, by its own weight. In contrast to this, we shall see that in order to push an ophiolite onto a continent, very special circumstances like the encounter between very young oceanic lithosphere and a zone of forced continental subduction, must prevail.

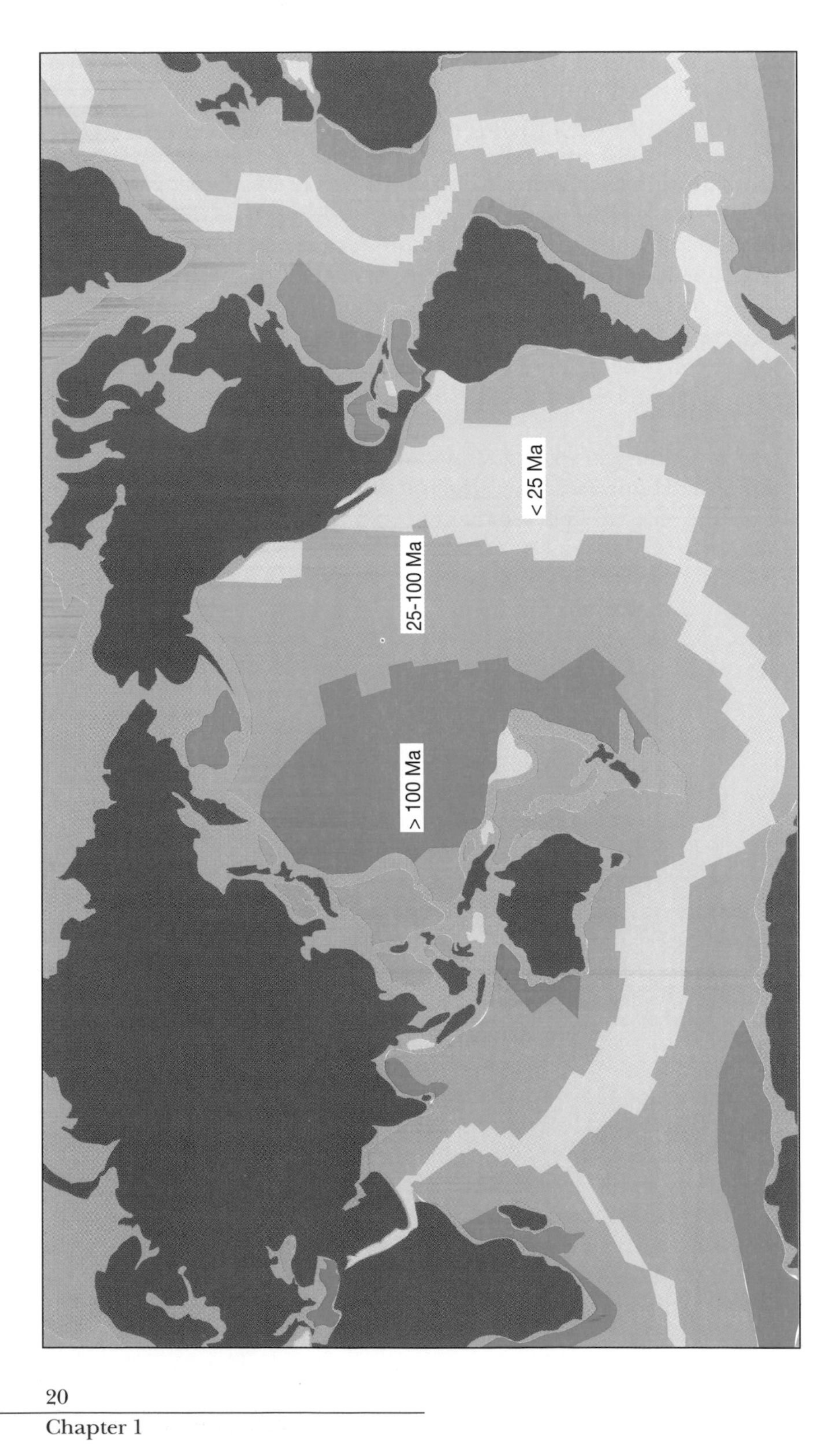
< 25 Ma
25-100 Ma
> 100 Ma

◀ **Figure 1.12**
Age of the Pacific lithosphere, showing younger ages to the east responsible for forced subduction along the North and South American Cordilleras, as opposed to the older ages to the west giving rise to spontaneous subduction along the Marianas and Tonga trenches. In the oceans, the blue shading increases in intensity with age. Continental margins and island arcs are shown in *light brown.* (After P. Molnar and T. Atwater 1978. Earth Planet. Sci. Lett., 41, 330–340)

2 The Earth – a Heat Engine

The terrestrial mantle is continuously pervaded by slow movements (convection) which transfer the internal heat of the Earth to its surface and reheat the cold plates plunging along the subduction zones, which thus play a key role in the control of this circulation. The oceanic ridges are not necessarily located right above the zones of ascent of these convection cells. The hotspots, areas of particularly strong volcanism, are fed by jets of hot mantle, the so-called plumes, in which rapid upward movements account for an important part of the heat transfer to the surface. Hotspots are located either on ridges, e. g. the Iceland ridge or within plates, e. g. Hawaii.

The ascent of hot material along the ridges, the drift of the various plates, and plunging of cold lithosphere along subduction zones are all parts of a closed circuit of material which pervades the mantle. Geochemists have actually now identified in the material rising along the ridges noble gases like xenon, krypton and helium, the isotopic composition of which indicates their sedimentary origin. They thus must have travelled, together with their enclosing sediments, through a subduction zone, thereby undertaking a complete round-trip at least once.

What do we know about these circuits in the mantle? and about the potential shortcuts, the hotspots, volcanoes through which a considerable volume of basalt escapes to the surface? The highest relative relief on our planet is not represented by the 8848 m by which Mt. Everest rises above sea level but by a volcano, the island of Hawaii, which at a peak elevation of 4200 m above sea level rises altogether about 10000 m above the floor of the Pacific Ocean, which in this region is deeper than 5000 m. The highest mountain known so far in our solar system is another volcano, Mt. Olympus on Mars, which

Figure 2.1
Mount Olympus on Mars

towers about 25 km above the surrounding plains (Fig. 2.1). Both mountains represent hotspots. However, what is the driving force of the brewing mantle? Without it, the Earth would be made up of a single plate like Mercury or the Moon. The reliefs worn down by erosion would not be regenerated by tectonic activity between different plates and the planet would be under a uniform cover of water, which would be pierced only locally by volcanic islands, the sole refuges supporting aerial life.

Convection in the Mantle

It is now generally accepted that the mantle is subjected to convection movements. When saying convection, we imply mobile and deformable material, and thus temperatures sufficiently high for the mantle to become deformable or plastic. These movements are slow, as they take place in the solid mantle and their velocity is in the same range as that of the plates (1–10 cm/a) which represent their expression on surface. As in the heat engine driven by steam or petrol, the movement is caused by the great difference in temperature between the interior of the engine and the exterior. Thus the temperature at the mantle/core interface is 3000–3500 °C (Fig. 2.6) compared to a mean surface temperature of the Earth of 15 °C. The lithosphere, which is denser than its underlying mantle because of thermal contraction (cf. Chap. 1) tends to descend and the deeper, warmer, and lighter layers tend to rise. This will lead to a convection cell, the functioning of which we may observe when we slowly heat a pot of water on a fire. The water layers in contact with the hot floor of the container expand and become lighter than the higher layers, and eventually change place with them.

The Convection Circuits

Numerous questions are presently still being debated concerning the circuits induced by convection in the mantle, such as the shape of the convection cells and the nature of the movements which keep them going. This subject is gaining increasing attention as the process has been shown in operation by means of cogent and spectacular geophysical evidence. We refer here to laser telemetry, which allows us to determine by means of laser beams sent from satellites the position of individual plates with a precision of a few millimetres and, concurrently, to trace their present movements with equal precision. Anomalies of the geoid reflecting a surplus or deficit of mass in the Earth's interior are reconstructed from modifications in the orbit of certain satellites and from measurements of the ocean surface carried out by these satellites. Seismic tomography allows us to trace with the aid of seismic waves the distribution of hot and cool zones within the mantle (Fig. 2.2), just as medical tomography delineates a tumour in healthy tissue by the temperature difference it creates.

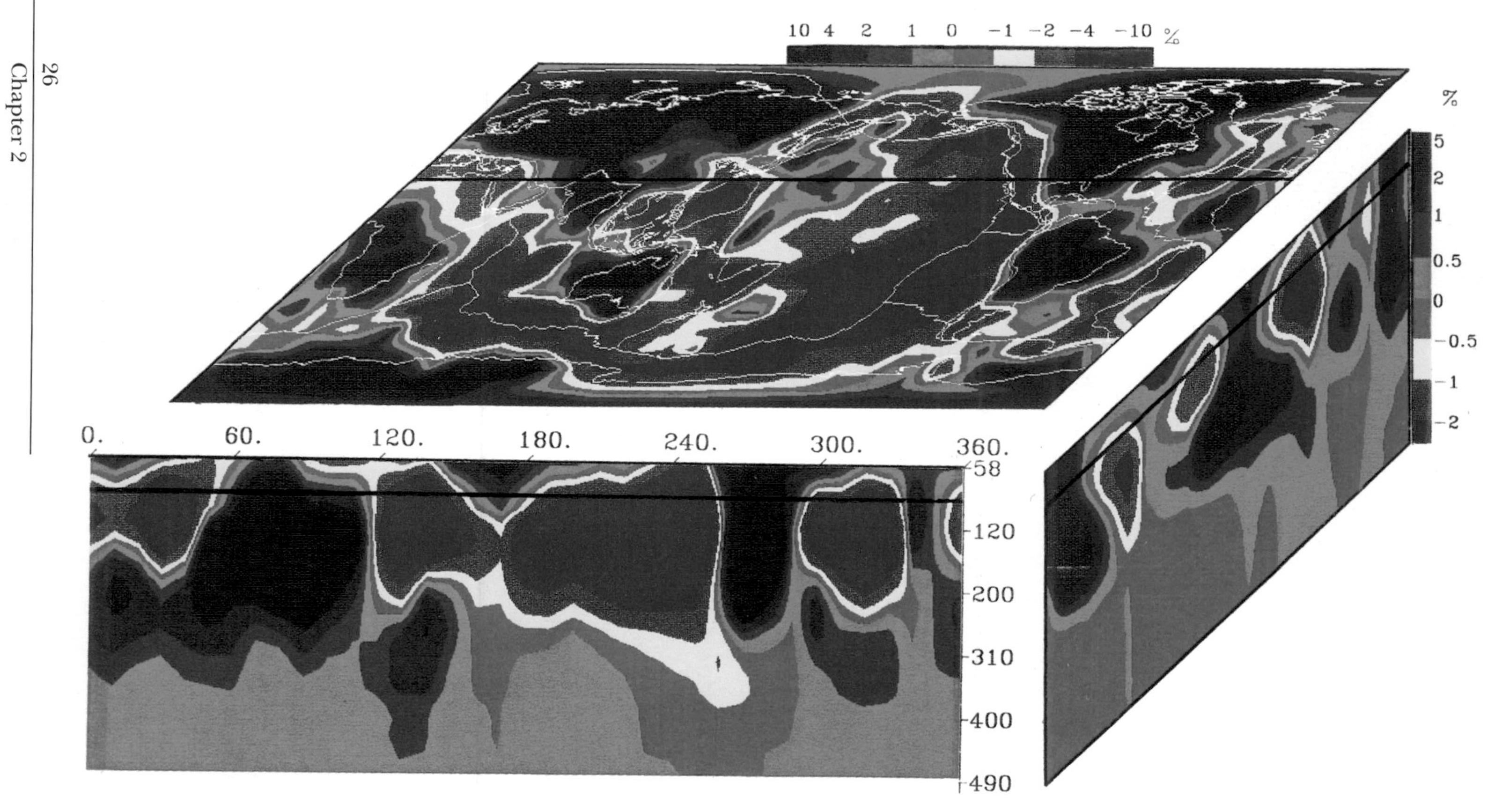
10 4 2 1 0 −1 −2 −4 −10 %
%
5 2 1 0.5 0 −0.5 −1 −2
0. 60. 120. 180. 240. 300. 360.
58 120 200 310 400 490

◀ **Figure 2.2**
Seismic tomography of the Earth. Map projection of the situation at -100 km, east-west section *along the solid line* in the map (*front*), and north-south section *along the side* of the map. The *colours* represent variations in the velocity of seismic waves which are interpreted as temperature differences. Hot zones of slow propagation are shown in *red*, and cold zones of fast propagation in *blue*. The hot zones correspond to the ridges and their roots, i. e. young oceanic mantle. The blue zones represent old oceanic lithosphere, continental lithosphere, and subduction zones. (J. P. Montagner)

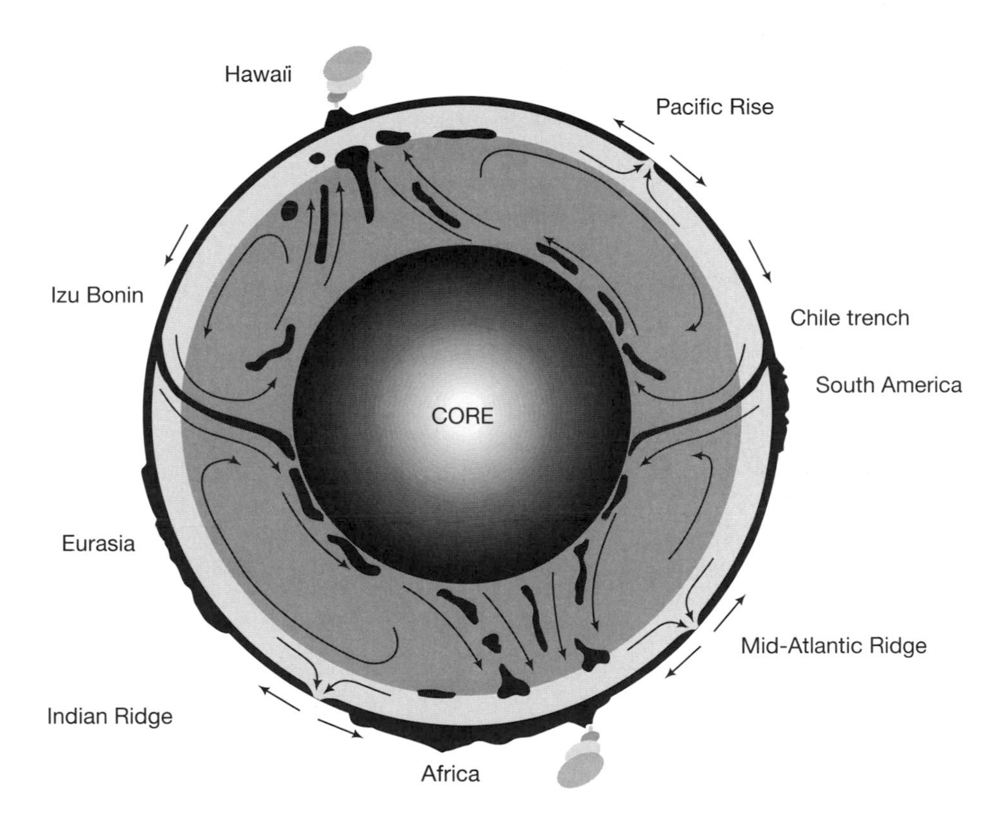

Figure 2.3
Illustration of a possible convective circulation involving the entire mantle. The plates (in *black*) descend along the subduction zones towards the core, traverse the upper mantle (*light grey*) and disperse prior to becoming resorbed in the lower mantle (*dark grey*). Their great longevity (100 Ma) allows them to rise again under the hotspots, contributing eventually to the volcanism at these points. The ridges are fed mainly by material ascending in the upper mantle. (After P. G. Silver and R. W. Carlson 1988. Ann. Rev. Earth Planet. Sci. , 16, 477–541)

We know by now that ridges and subduction zones do not play a symmetric role in convection. Lithosphere plunging into subduction zones clearly marks the descending part of a convection cell, but a ridge does not necessarily coincide with its ascending portion (Fig. 2.3).

This asymmetry is easily explained. Descending lithosphere, about 100 km thick, requires some 100 Ma to disappear due to heating and subsequent melting in the hot mantle. This duration results from the law of thermal diffusion which here is taken as $5\sqrt{t}$ (in km and Ma) and not $9.5\sqrt{t}$ as in Chapter 1, because of the dependence of the diffusion coefficient with temperature. The thickness to be heated is 50 km, as the descending plate will be heated on either side. We thus arrive at $t = 100$ Ma. These 100 Ma represent a considerable time span over which a lithospheric plate, at the known velocities of 1–10 cm/a, may travel for 1000 to 10000 km. Due to their rigidity, the subducted plates may thus represent dams and "guiding surfaces" for convection currents even far away from their respective subduction zones (Fig. 2.3). One may also observe that these lithospheric plates descending into the hot mantle represent the main heat exchangers of the convective system or its radiators.

In contrast, we shall see that just a modest rise in asthenosphere will suffice to create a ridge: as soon as its depth becomes less than 100 km, melting occurs, and mantle bubbles detach from this melting mantle that rise rapidly to supply the ridge with basalts (Chap. 7). A ridge may thus possess a very shallow and accidental source like an upward flexure in a high-level convection system. We are dealing here with a situation which is different from ascending zones associated with hotspots (cf. later).

It is not yet known for certain whether convection bottoms out at 2900 km at the core-mantle interface (Fig. 2.3) or at 670 km, where a sudden rise in seismic velocities takes place, thereby defining the boundary between upper and lower mantle. In the latter case, there would be a second convection level in the lower mantle (Fig. 2.4). It is true that the 670 km depth constitutes an important discontinuity at which, under the enormous pressure exerted by the weight of the overlying rocks, the last silicates as we know them from the surface will disappear. New mineral species will form from them which are denser and possess physical properties which can resist the descent of upper mantle material below this level. The deepest earthquake sources tied to subduction zones are located at a depth of around 700

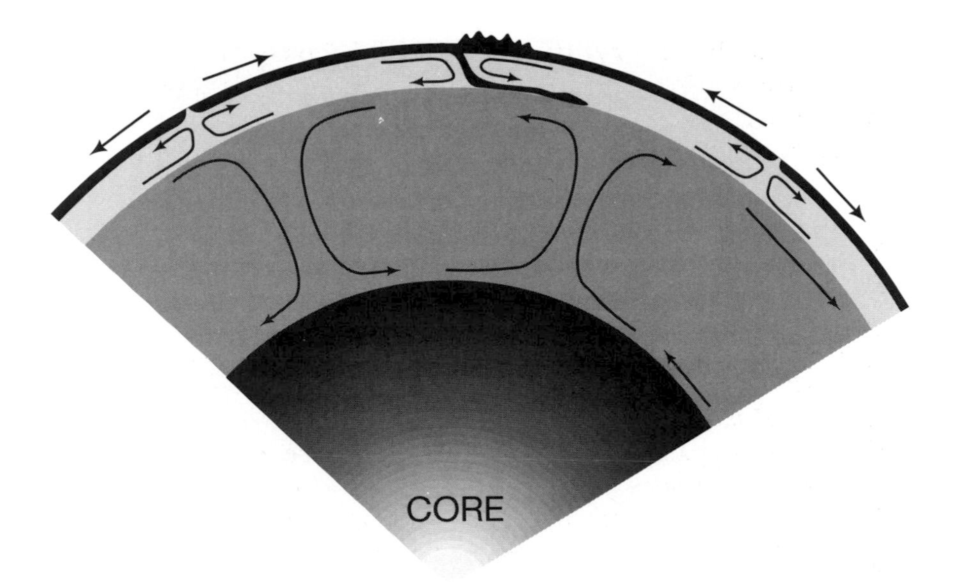

Figure 2.4
Illustration of a convective circulation in two levels in the mantle. The elongated cells of the upper mantle descend along subduction zones and rise below ridges. The circulation in the lower mantle may be mechanically coupled with that of the upper mantle (same direction in horizontal movements, opposite direction in vertical movements), thermally coupled (same sense in vertical movements, opposite sense in horizontal movements) as rather suggested by seismic tomography, or not coupled at all

km. This might suggest that the lithosphere, whose fracturing is the underlying cause of earthquakes, does not penetrate beyond this depth. Geochemists also favour convection on two levels, as the ridges' basalts and those from hotspots possess mineralogical characteristics indicative of provenance from different reservoirs. Convection on two levels might be the underlying cause of these different reservoirs. However, certain images derived from seismic tomography suggest that the lithosphere of subduction zones locally transgresses the 670 km discontinuity and penetrates into the lower mantle, whereas elsewhere it would creep along this discontinuity (Fig. 2.5). This could serve as an argument for the existence of a single convection level in the mantle. Figure 2.3 attempts a reconciliation of these contradictory data within the framework of a single-level convection system with the possibility of local instabilities above 670 km which would be responsible for the ridge activities. The zones of ascent of the deeper convection would be situated below the hotspots, which we shall examine later. Accumulation of dense litho-

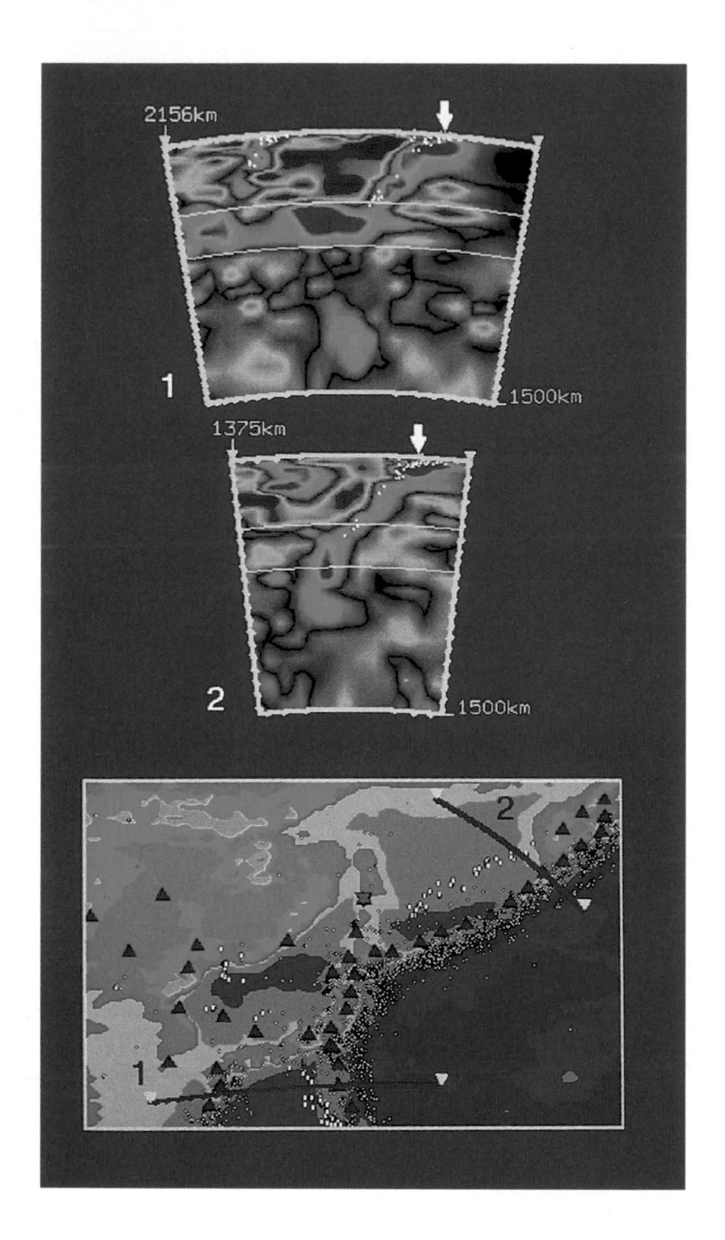

Figure 2.5
Lithospheric slabs plunging into subduction zones can either be deflected by the 670 km discontinuity and slide along this surface (cross section 1) or cross this discontinuity and penetrate into the lower mantle (cross section 2). This is shown by two tomographic sections across the Bonin-Ryuku subduction zones, east of Japan and Kuriles. The code for colours is the same as in Fig. 2.2. The colder and seismically faster slabs appear in *blue* and *green*, contrasting with the hotter and slower surrounding asthenosphere in *yellow* and *red*. *White dots* represent location of earthquakes tracing local ruptures in the descending slab. The *yellow lines* correspond to the 400 km and the 670 km mantle discontinuities. (Van der Hilst et al., Nature, 353, 37–43; modified by the author)

spheric slabs above the 670 km discontinuity as shown in Fig. 2.5 may trigger rapid avalanches of slabs into the lower mantle. Such turnovers would have major consequences on mantle convection.

Taken a mean mantle circulation velocity of 5 cm/a, comparable to the drift velocity of the plates, we arrive at a time span of 200–400 Ma for the material to undertake a complete roundtrip in a convection cell if we assume that the surface extent of the cell is 5000 or 10000 km respectively. This admittedly rough estimation nevertheless leads to the conclusion that since the origin of the Earth some 4500 Ma ago, the mantle has undergone considerable mixing. The periods of 200–400 Ma we shall encounter again in our investigations of major cycles marking Earth history (Chap. 9).

Geothermal Gradients and Convection

The heat generated in the Earth's interior dissipates either by conduction, i. e. by diffusion through static material like the lithosphere, or by convection, in which case it is transported by material in motion. For "long-distance" transport, convection is much more efficient than conduction, taking into account its assumed velocities of 1–10 cm/a. As we have seen, it takes about 100 Ma for heat to traverse a 50 km slab thickness by conduction, whereas over the same span of time it is conveyed by convection over 1000 to 10000 km. Within the Earth, superimposed convection cells are separated by relatively stagnant envelopes (otherwise the cells would mix with each other) within which thermal conduction dominates. These zones are referred to as boundary layers. The change of temperature with depth, the so-called geothermal gradient, is an expression of this situation. The geothermal gradient in the boundary layers is also higher, as the diffusion of heat here is more difficult. It is admitted that within the convection cells the exchange of heat is rather limited because of the higher efficiency of convection compared to that of conduction. In thermodynamics such an environment is referred to as adiabatic. In an adiabatic gradient the slope is 10 to 100 times less steep than in a conductive gradient. The existence of such a gradient results from the fact that temperature is on our scale the expression of the physical interactions between atoms which increase with pressure, bringing the atoms into closer contact with each other, thus raising the temperature. This relationship between temperature and pressure in an adiabatic system is put to work in the refrigerator: a gas cools during adiabatic decompression and the cold is

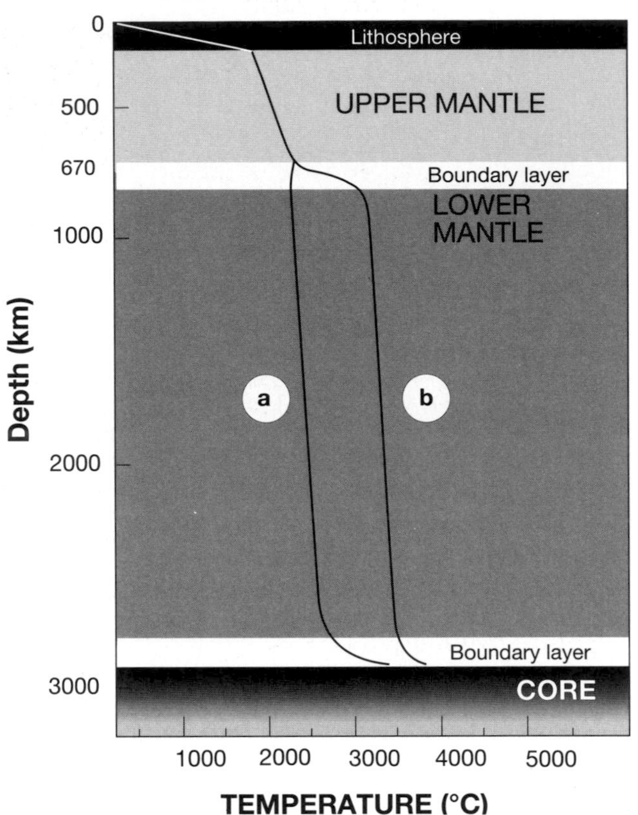

Figure 2.6
Increase of temperature within the Earth as a function of depth. The curve, or geotherm, permits an estimate of the temperature at a given depth. Starting from the surface: the lithosphere (in *black*) is a boundary layer within which the temperature increases rapidly with depth (high geothermal gradient). For convection at a single level (case *a*), temperature increases following a low adiabatic gradient until the mantle-core boundary layer is met. For convection in two levels (case *b*), the temperature is much higher in the lower mantle because of the boundary layer at 670 km between the upper and lower mantle. (After P. G. Silver and R. W. Carlson 1988. Ann. Rev. Earth Planet Sci., 16, 477–541)

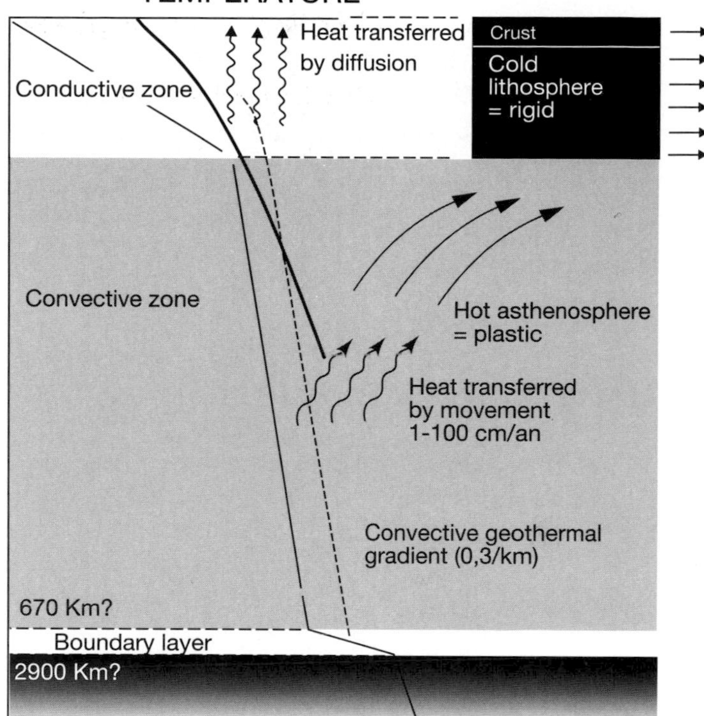

Figure 2.7
Simplified diagram of temperature vs. depth (cf. Fig. 2.6) illustrating the relationship between heat transfer and mechanical structure within the mantle. The heat is transferred by diffusion (conduction) in the rigid lithosphere and the boundary layers at depth and by movement of material (convection) in the convection cells. The different slopes of the geotherms (cf. Fig. 2.6) correspond to the different types of transfer. The *thicker line* (solidus) corresponds to the onset of melting in the mantle (cf. Chap. 6). The *dashed line* illustrates the path of a plume rising from a deep boundary layer. It intersects the solidus at great depth, which may explain the abundance of melting (Chap. 6)

then passed by conduction to the interior of the refrigerator. This also explains why a bicycle pump heats up when used: rapid adiabatic compression of the enclosed air takes place, leading to a rise in temperature. This heat is then transmitted to the body of the pump.

From the surface inwards, the temperature may rise in a step-like pattern along the gradients of Fig. 2.6, which are idealized in Fig. 2.7. The first boundary layer is the lithosphere, separating the convection of the mantle from that of the oceans or the atmosphere. Then comes a cell which, depending on the authors, stops at a depth of 670 or 2900 km. A boundary layer with a strong conductive gradient separates this cell from the next cell situated either in the lower mantle or in the core. We must keep in mind that the terrestrial magnetism is proof of convection taking place in the mantle. It is induced, as in a generator, by the movement of the conductive metallic alloys making up the core.

Hotspots and Mantle Plumes

In the introduction to this chapter, we mentioned that the points of most pronounced relief on Earth and in our solar system are both volcanic hotspots, viz Hawaii and, respectively, Mount Olympus on Mars. They are hot because the outflow of lavas and the related supply of heat from depth here are considerable. The height of the relief is the consequence of the massive supply of basalt and of regional thermal dilatation on the scale of at least a 100 km. It is generally conceded that this exceptional activity results from the ascent, under the hotspot, of a jet of hot mantle charged with basaltic liquid, a so-called "plume" (Fig. 2.8). Active volcanism on Hawaii extends laterally over 50–100 km, suggestive of a similar diameter for the plume feeding it. The ascent velocity in the mantle will be in the range of several tens of centimetres per year. Hotspots are rather numerous and may be either oceanic like Hawaii, Tahiti, Iceland, The Canary Islands and Réunion, or continental like Yellowstone or the Afars, south of the Red Sea (Fig. 2.9).

In oceanic environment we may distinguish hotspots within plates like Hawaii from those located on a ridge, i. e. between two diverging plates as in Iceland. The activity of the latter hotspots contributes to oceanic spreading. Thus spreading along the Mid-Atlantic Ridge is supported, from north to south, by hotspot volcanism on Iceland, the Azores, Ascension Island, Tristan da Cunha and Bouvet Island (Fig. 2.9). In contrast, the plumes feeding within plate hotspots pierce these plates, as a blow torch cuts through a piece of sheet metal. The volcanic island of Hawaii represents the southeasternmost of a chain of islands which increase in age and degree of erosion towards WNW (Fig. 2.10). The chain extends farther below sea level in the form of a submarine chain of guyots which represent ancient volcanoes, the flat tops of which are evidence of subaerial erosion preceding their sinking below sea level. This submergence

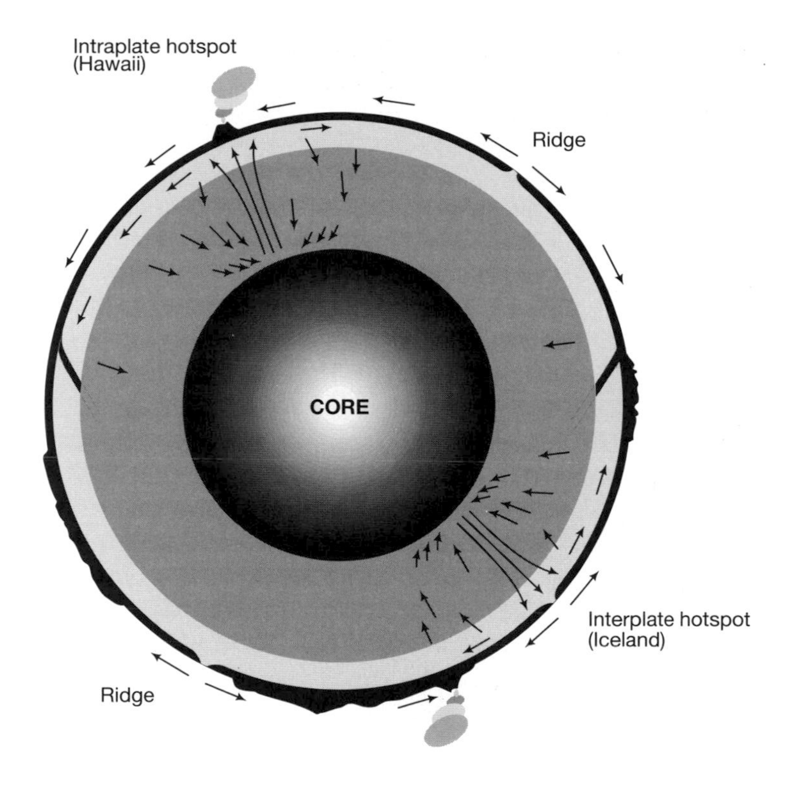

Figure 2.8
Plumes and hotspots. In this schematic section of the Earth, plumes rise from the mantle-core interface and make up the jets of hot mantle material feeding the hotspots. The Hawaii plume penetrates a plate which slides above it, whereas the Iceland plume located below the Mid-Atlantic Ridge contributes to the separation of the European and American plates. (After V. Courtillot 1973, La Recherche, 32, 270–272)

results from thermal contraction and subsidence of the lithosphere receding from the hotspot. The oldest volcanoes of the Hawaii chain are 40 Ma in age at the turning point of the trend in Fig. 2.9. From this point to the NW the Emperor chain continues, with the oldest guyots at the point of plunging into the Aleutian subduction zone attaining an age of 80 Ma. These trends indicate that the Pacific plate changed its position relative to the hotspot (Fig. 2.11), or to come back to our example, that the metal plate shifted its position above the blow torch. We may observe further in Fig. 2.9 that the other hotspots of the Pacific plate exhibit pronounced similarities in trends. These are the hotspots of Macdonald, Sala y Gomez and Juan de

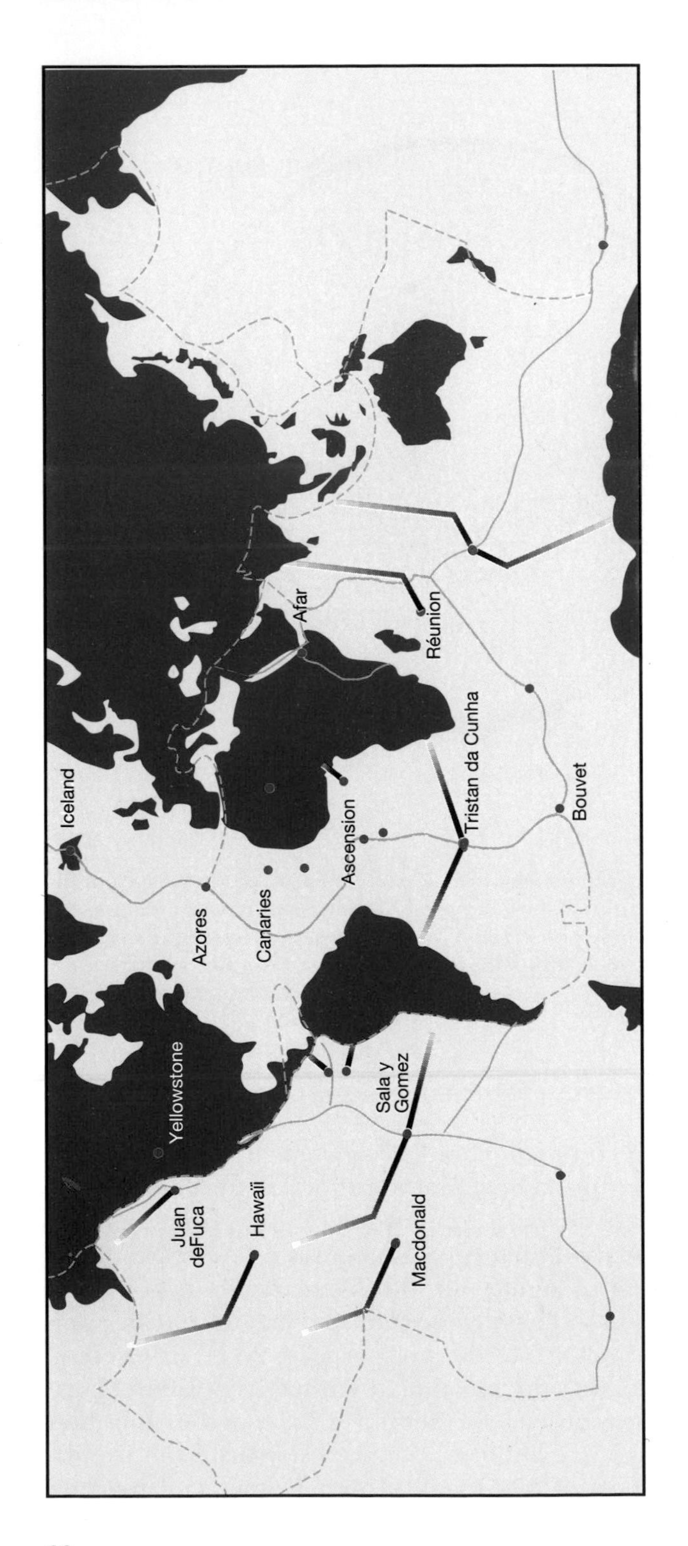
Iceland
Azores
Canaries
Ascension
Tristan da Cunha
Bouvet
Afar
Réunion
Yellowstone
Juan deFuca
Hawaii
Sala y Gomez
Macdonald

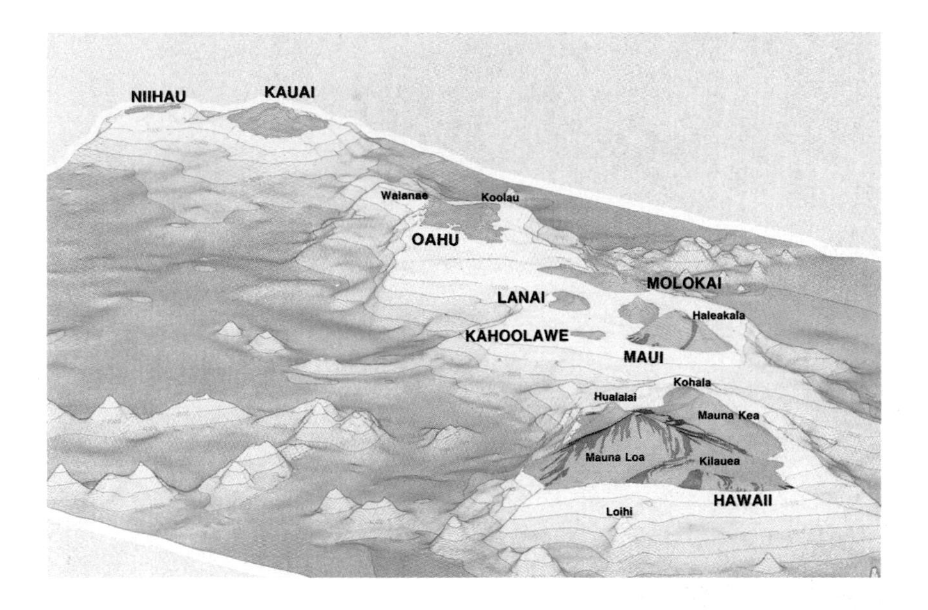

Figure 2.10
Topographic relief of the islands of the Hawaii hotspot.Today, the hotspot itself is located in the eastern part, ahead of the island of Hawaii. Towards the northwest (*left back* of the drawing), we move away from the hotspot, as illustrated by more intense erosion and lower relief. (R. I. Tilling et al. 1987, Eruptions of Hawaiian volcanoes. U. S. Geol. Survey Ed, 54 p.)

◀ **Figure 2.9**
Main hotspots (*red dots*) on the surface of the globe in relation to the plate structure. The hotspots punctuate the ridges (inter-plate hotspots) as in Iceland or the Azores, or pierce the plates themselves (intra-plate hotspots) as in Hawaii. The dark grey lines represent the traces left on the plates by the previous activity of the hotspots in the form of rises or chains of submarine volcanoes. Note the parallel traces of the hotspots of Hawaii, MacDonald and Sala y Gomez on the Pacific plate. From this, one may reconstruct the drift of this plate in relation to the hotspots considered as fixed. (After V. Courtillot 1973, La Recherche, 32, 270–272)

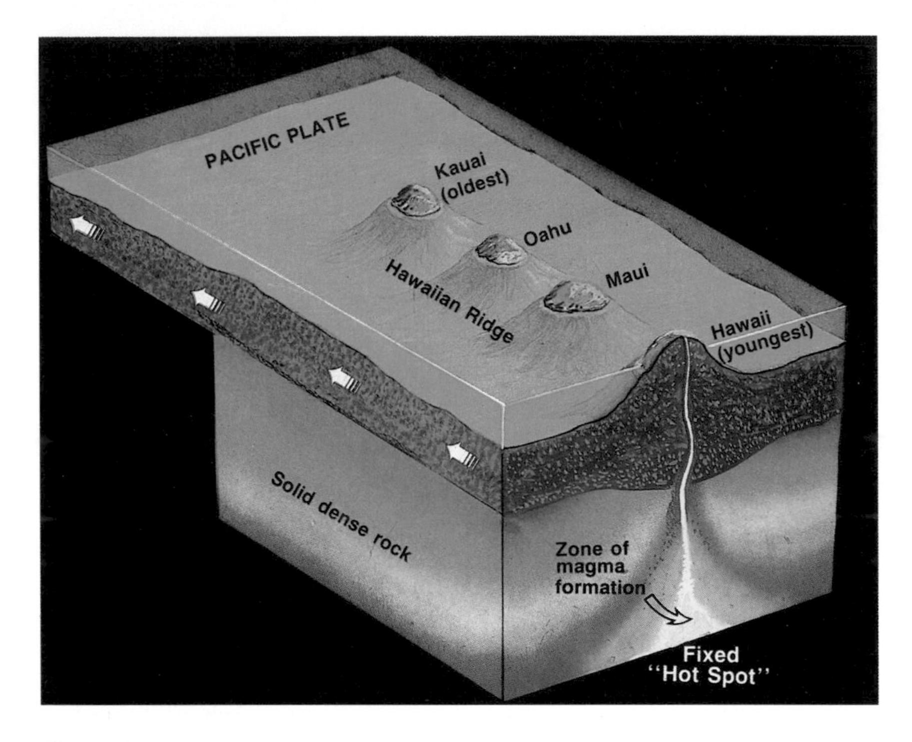

Figure 2.11
Schematic presentation of the drift of the Pacific plate over the Hawaii hotspot. (R. I.Tilling et al 1987, Eruptions of Hawaiian volcanoes. U. S. Geol. Survey ed., 54 p.)

Fuca. This similarity shows that the four "blow torches" have remained more or less stationary relative to each other while the Pacific plate shifted above them. This is a highly significant observation as it permits us to reconstruct absolute, and not only relative, plate movements. Thus, over the presumably fixed reference system presented by the hotspots we may deduce from Fig. 2.9 backward in time, that the Pacific plate experienced a regular drift towards the WNW over the last 40 Ma, preceded by a drift to the NNW between 40–80 Ma ago.

The building-up of a 25-km-high volcano on Mars and the absence of trends on the surrounding plains indicate that the lithosphere of this planet is fixed relative to the hotspot. The possibility of accumulating such a relief proves that there are no plate tectonics on Mars.

The Chemical Fingerprints of Hotspots

The study of the geochemistry of the basalts emanating from the hotspots exhibits compositional differences with basalts extruded on ridges farther away from hotspots. These ridge basalts are referred to by the acronym MORB for Mid-Oceanic Ridge Basalts. At first sight, the concentrations of the major elements are not of discriminatory value. The elements Si, Al, Fe, Mg, K and Ti, which as oxides make up 99 % of the terrestrial rocks, exhibit only subtle differences. Only potassium and titanium tend to be present at slightly higher rates in hotspot basalts.

A more detailed evaluation of the geochemistry of the trace elements which are present in basalts in concentrations below 0.03 % have become possible since the early 1970s, when technical advances facilitated sufficiently high analytical precision. The elemental signatures of basalts from ridges and from hotspots are different indeed. Hotspots basalts are higher than ridge basalts in elements like K, Rb, P, Ti and LREE. This shows that the source of the basalt plumes is a mantle reservoir less depleted in these elements by successive phases of melting than the reservoir feeding the ridge basalts.

Isotope geochemistry, as briefly outlined below, permits us to penetrate to the very heart of the elements. The atoms of a certain element may be distinguished by slightly different masses because of differences in the number of neutrons in their nuclei. Such atoms are referred to as isotopes. With the introduction of mass spectrometres in laboratories, it became possible to separate the various isotopes of an element and to use the isotopes and their mutual ratios either as tracers of different source environments or as time recorders, as some of them disintegrate at constant rates. Isotope geochemistry has confirmed the enriched nature of the hotspot reservoir and furthermore led to the conclusion that the two reservoirs stopped communicating about 2 Ga ago, after which they embarked on independent evolutionary paths. At present, mixing phenomena between components derived from these distinct reservoirs may be observed along the oceanic ridges.

Isotope Geochemistry and Mantle Reservoirs

Let us recall that an atom is made up of its nucleus and negatively charged electrons orbiting around it, their number equalling that of the positively charged protons in the nucleus. In the latter, there are also neutral particles of the same mass as the individual protons, i. e. the uncharged neutrons. One now defines an element by its atomic number Z (number of electrons or protons) and the isotopes by their mass number A (fixed number of protons and variable number of neutrons in the nucleus). We thus write

${}^{A}_{Z}X$ or e.g. ${}^{204}_{82}Pb$, ${}^{206}_{82}Pb$, ${}^{207}_{82}Pb$, ${}^{208}_{82}Pb$, for natural lead isotopes.

The elements are arranged with increasing mass from the lightest, hydrogen with only one proton, to the heaviest natural element, uranium with 92 protons in the nucleus. The hydrogen nucleus is usually made up of only one proton. When it contains additionally one neutron, we are dealing with the isotope deuterium. Whereas the isotopes of lighter elements like hydrogen, carbon or oxygen usually are stable, this does not apply to certain isotopes of heavier elements which decay with time into so-called "radiogenic" isotopes of a different element. Strontium (Z = 38), for instance, is made up of three stable isotopes (^{88}Sr, ^{86}Sr, ^{84}Sr) in near-constant relative abundances and one radiogenic isotope (^{87}Sr) derived from the radioactive decay of the rubidium isotope ^{87}Rb isotope (Z = 37). The abundance of ^{87}Sr in rocks depends on the original rubidium content of the rock and the time elapsed since the formation of the rock. The nuclear decay of a certain isotope takes place at a constant rate, the latter, however, differing for each natural radioactive source isotope. It is generally held that measuring the ratios between source and daughter isotopes and the ratio between the radiogenic isotope and its respective stable sister isotopes will facilitate the identification of an environment and of the time the rock has remained in it.

For our problem, isotope geochemistry in particular demonstrates that the ridge basalts possess a mean value of $^{87}Sr/^{86}Sr$ = 0.7027 whereas in hotspot basalts this is 0.704. As a consequence, it is concluded that the sources of these two types of basalts are different. The hotspot basalts thus should exhibit a higher Rb/Sr than the MORB because the decay of ^{87}Rb into ^{87}Sr is responsible for increasing this radiogenic isotope in the mantle reservoir. One may conclude furthermore that the separation of the two reservoirs is an ancient feature which became established close to 2 Ga ago.

One of the fundamental findings of isotope geochemistry, furnished by the study of the Sr isotopes, is that the soure of the oceanic basalts is generally depleted with respect to the conceptual composition of the primitive mantle. This very ancient depletion is the result of repeated phases of melting which have led to the formation of the continental crust, especially over the first 2 Ga of the history of our planet.

What is the origin of these mantle plumes? Why are they stationary? One part of an answer is furnished by the hypothesis which interprets the plumes as ascending from a boundary layer at the base of the convection cells (Fig. 2.7), their ascent either participating in that of the convection cells themselves or being independent of them. The connection to a boundary layer, either at the 670 km discontinuity for double-level convection or at the core-mantle interface for single-level convection, would lead to a fixed position with respect to the upper mantle shell from which the ridges draw their material. The geochemical signature of the hotspot basalts would reflect this deep source. A more in-depth discussion of these still very much debated questions presupposes an incursion into the mantle along an even more difficult pathway.

The Origin of Hotspots

The origin of hotspots in the mantle is the object of two differing hypotheses. Although they might differ at present, it cannot be excluded that with the progress of our knowledge about convection in core and mantle and about exchanges between these two shells, a connection between these two hypotheses might show up. The hypothesis tying the origin of the hotspots to the existence of the thermal shield of Pangea will be discussed at the end of this book in Chapter 9. Let us simply say that the hotspots result from the accumulation of heat below thick continental lithosphere, the accumulation itself being the result of the poor thermal conductivity of this thick lithosphere which acts as a thermal shield.

In the second hypothesis, the plume responsible for a hotspot is detached from a boundary layer. Figure 2.7 illustrates that in the simplified system considered here this would be the only possible means to bring more heat and more basalt to the surface than at the ridges. We indeed have to admit that mantle rising via a convection cell will follow an adiabatic gradient and thus will reach the base of the lithosphere at a temperature which is independent of its depth of origin from within this convection cell. In contrast to this, the thickness of the lithosphere intervenes directly in the melting, as we shall see in Chapter 6. To arrive at the top of the asthenosphere at a temperature that is above the surrounding temperature, the source has to be located in the boundary layer below the cell and the plume has to rise sufficiently fast to retain its heat with respect to the surrounding mantle (Fig. 2.7). The rate of ascent will be in the order of several decimetres per year, which makes any thermal exchange with the vicinity negligible. We shall now demonstrate how we can estimate this rate of ascent of a plume. The simplified calculation is carried out by dividing the mantle material flux by the surface area of the plume. The mantle flux itself is derived from the mean extrusion rate of lavas on surface through the hotspot and the plan area of the plume is taken as the area of active volcanism of the hotspot. For Hawaii this is

$$\pi \frac{50^2}{4}\ \mathrm{km}^2 = 1.963\ \mathrm{km}^2.$$

The flux of mantle material is estimated from the volcanic productivity of Hawaii (0.1–0.2 km^3/a) and from a 10 % mean extraction rate of magma of the mantle. This leads to a mantle supply rate of 10 x 0.1 km^3/a or 1 km^3/a. The velocity of mantle ascent, which is the mantle flux divided by the hotspot area, is then about 50 cm/a.

3 The Ridges – Cradles of the Oceans

From ridges spreading at rates close to 20 cm/a, to continental rifts opening at only a few millimetres per year, the total length of the expansion belt spans some 75000 km. What do we know about the oceanic ridges and how do we know it? It is primarily the diagnostic shape of their relief which allows us to distinguish between different types of ridges on the basis of their spreading velocities, high or low. The ridges are by no means uninterrupted bands. They are cut into segments by faults which displace them and in the case of fast-spreading ridges by strange terminations, the so-called overlapping spreading centres. It was found as a surprising discovery from satellite data that the submarine relief is also discernible from highs and lows on the surface of the ocean.

Submarine Topography and Geology

A strange candy-coloured picture appeared recently on the cover of the scientific journal Nature (1988). We have reproduced it here together with another figure from the corresponding article (Fig. 3.1). They represent the submarine topography of the equatorial zone of the East Pacific Rise. The strange appearance of the figures results from the considerable vertical exaggeration of the relief and from the use of a wide spectrum of colours to illustrate the successive depth levels. The floors are shown in blue whereas mounds and small submarine hills, which appear as unscalable peaks, are shown by reddish colours, changing upwards through orange and yellow into white. The recording of such maps along oceanic ridges requires a dense coverage of the region studies by the oceanographic vessel. Such vessels are equipped with a highly sophisticated probe, the SEA-

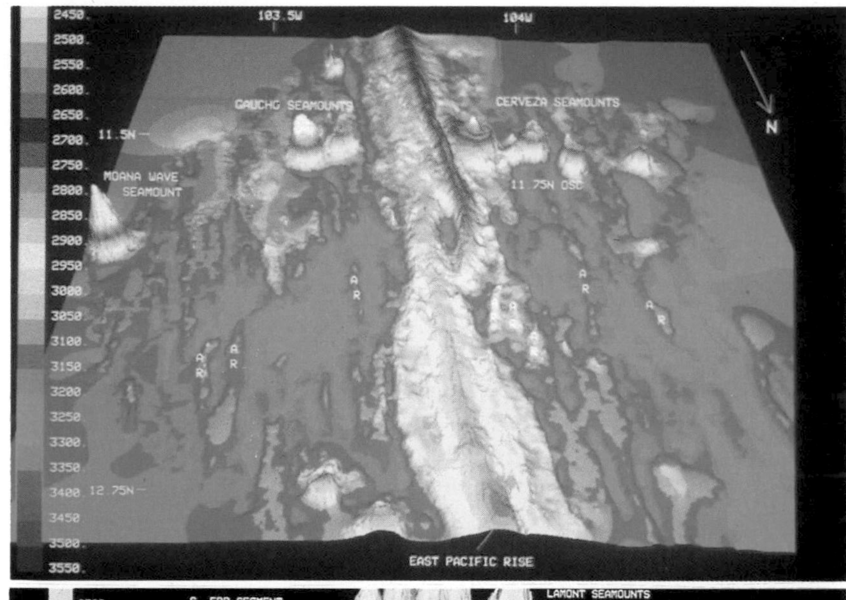

2450.
2500.
2550.
2600.
2650.
2700.
2750.
2800.
2850.
2900.
2950.
3000.
3050.
3100.
3150.
3200.
3250.
3300.
3350.
3400.
3450.
3500.
3550.
103.5W
104W
GAUCHG SEAMOUNTS
CERVEZA SEAMOUNTS
N
11.5N
MOANA WAVE SEAMOUNT
11.75N OSC
A R
12.75N
EAST PACIFIC RISE

a

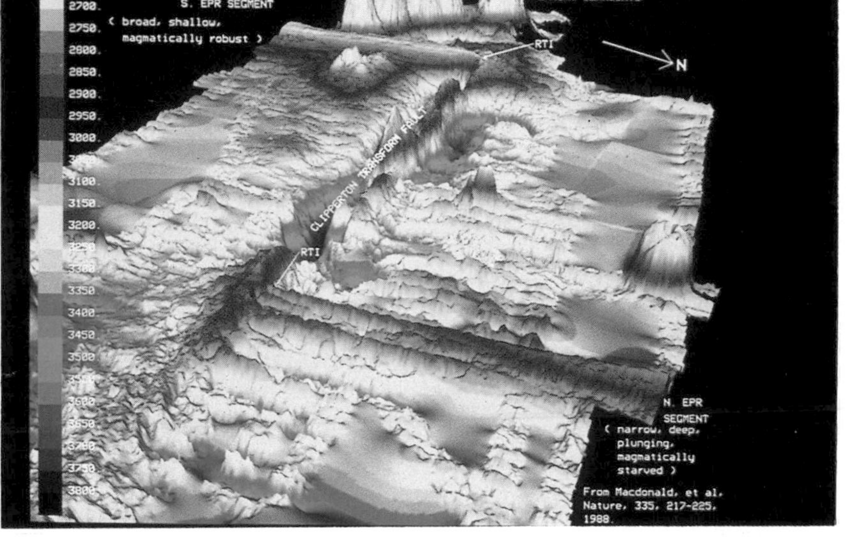

2700.
2750.
2800.
2850.
2900.
2950.
3000.
3100.
3150.
3200.
3350.
3400.
3450.
3500.
3600.
3650.
3800.
S. EPR SEGMENT
(broad, shallow, magmatically robust)
LAMONT SEAMOUNTS
RTI
N
CLIPPERTON TRANSFORM FAULT
RTI
N. EPR SEGMENT
(narrow, deep, plunging, magmatically starved)
From Macdonald, et al. Nature, 335, 217-225, 1988.

b

◀ **Figure 3.1 a,b**
Maps of the East Pacific Rise. These colour-enhanced reliefs are highly exaggerated. (**a**) Hook-like faulting of two segments of the ridge (overlapping spreading centres) and the volcanic peaks (seamounts) extending from the rise on the two plates. They may result from an "excess" of activity in the feeding centre of the segment. (**b**) Two segments of the ridge separated by a fracture zone. The ridge farther back is highly active, as illustrated by its elevated relief (in *white*) and the volcanoes (seamounts) close to it. It tends to invade the transform domain, creating a relief between two basins (in *blue*), the depth of which is more in keeping with a transform fault. (MacDonald et al. reproduced by permission of Nature 335, 217–225. Copyright © 1988, Macmillan Magazines Ltd.)

BEAM, which employs the sonar principle to explore the sea floor. This principle is well known from small pleasure boats. The sonar used in our case is highly perfected to fulfill a variety of tasks. The information received is processed by a computer and the relief highlighted by the SEABEAM is immediately dispatched on a paperchart following the advance of the vessel. The instrument has been perfected by the US Navy and we again have to admit that scientific oceanography owes much to the reconnaissance programmes carried out by this organization with the aim of being prepared for submarine warfare. Thus, one of the largest American oceanographic institutes, Scripps, is the scientific off-shoot of the large naval base at San Diego in southern California.

The SEABEAM was used for the first time for scientific purposes by the National Centre for the Exploitation of the Oceans (CNEXO), now the French Institute for Research and Studies of the Sea (IFREMER). French scientists thus were the first to draw an exact map of a part of the East Pacific Rise. It was discovered here that the zone of active volcanism is a straight, only 1–2-km-wide band filled with basalt, the great "Pacific Highway" (Fig. 3.1). These maps were used to guide to selected sites the small submersible CYANA which is able to undertake dives to a depth of 3000 m. It was named after the Greek maiden CYANA, the daughter of Scythias, who, together with her father, dived below the ships of the Persian fleet of Xerxes during the battle of Salamis in 480 B.C. to loosen their moorings and send them to the sea floor. With the discovery of the hydrothermal springs on the East Pacific Rise, made by CYANA and ALVIN, its American companion, what was to prove to be the finest submarine worksite, was opened with pomp.

Why should there be such a keen interest in submarine topography? Geologists know quite well that continental topography only poorly reflects the geometry of the underlying formations especially in areas with a somewhat complex tectonic situation. This is so because erosion levels off indifferently any relief which might be important from a geological point of view, highlighting more the contrasts in rock hardness which, nevertheless, are not without interest by themselves. On the ocean floor, however, erosion is virtually non-existent and the relief observed thus possesses immediate geological significance. We have already seen that beyond certain distance from the ridge, the mean elevation reflects thermal contraction of the lithosphere because of cooling. Differences in relief displayed by maps of the ridges (Figs. 3.1 and 3.4) generally are caused by the intrusion of hot material from depth or by fracturing of the crust along faults. The absence of covering sediments and surficial alteration close to the ridges resulting from the young age of the respective parts of the sea floor also aid these observations. Thus, a detailed topographic map of the sea floor over a ridge constitutes a document as rich in information as a map showing terrestrial geology. This is even more so if the nature of a characteristic relief is identified from dredged materials or by direct observation by means of manned or unmanned submersibles.

The Highs and Lows of the Ocean Surface

Satellite altimetry is carried out by measuring the altitude of a satellite above the ocean surface with the precision allowed by directed laser beams which, in the present case, have an error of a few centimetres over distances in the order of 1000 km. Assuming that the satellite is in a perfectly circular orbit and that the Earth is a perfect sphere, this distance should be constant. This is, however, not so! Even when we correct for the effects of the elliptic nature of the satellite orbit, the shape of the Earth in the form of a reference ellipsoid, and other effects like attraction of the satellite by terrestrial masses, tides and atmospheric or ocean currents, the distance between satellite and ocean surface will still change. This may be recognized in Figs. 3.2 and 3.3, especially in Fig. 3.2, which shows a low of 80 m in the Indian Ocean and a high of 100 m in the western Pacific when compared to the surface of the reference ellipsoid. The surface of the ocean thus is "bumpy" representing the so-called "geoid".

The highs and lows, which measure a few tens of metres in height, extend laterally over several thousand kilometres. The smallest ones measurable, several decimetres in height, cover dis-

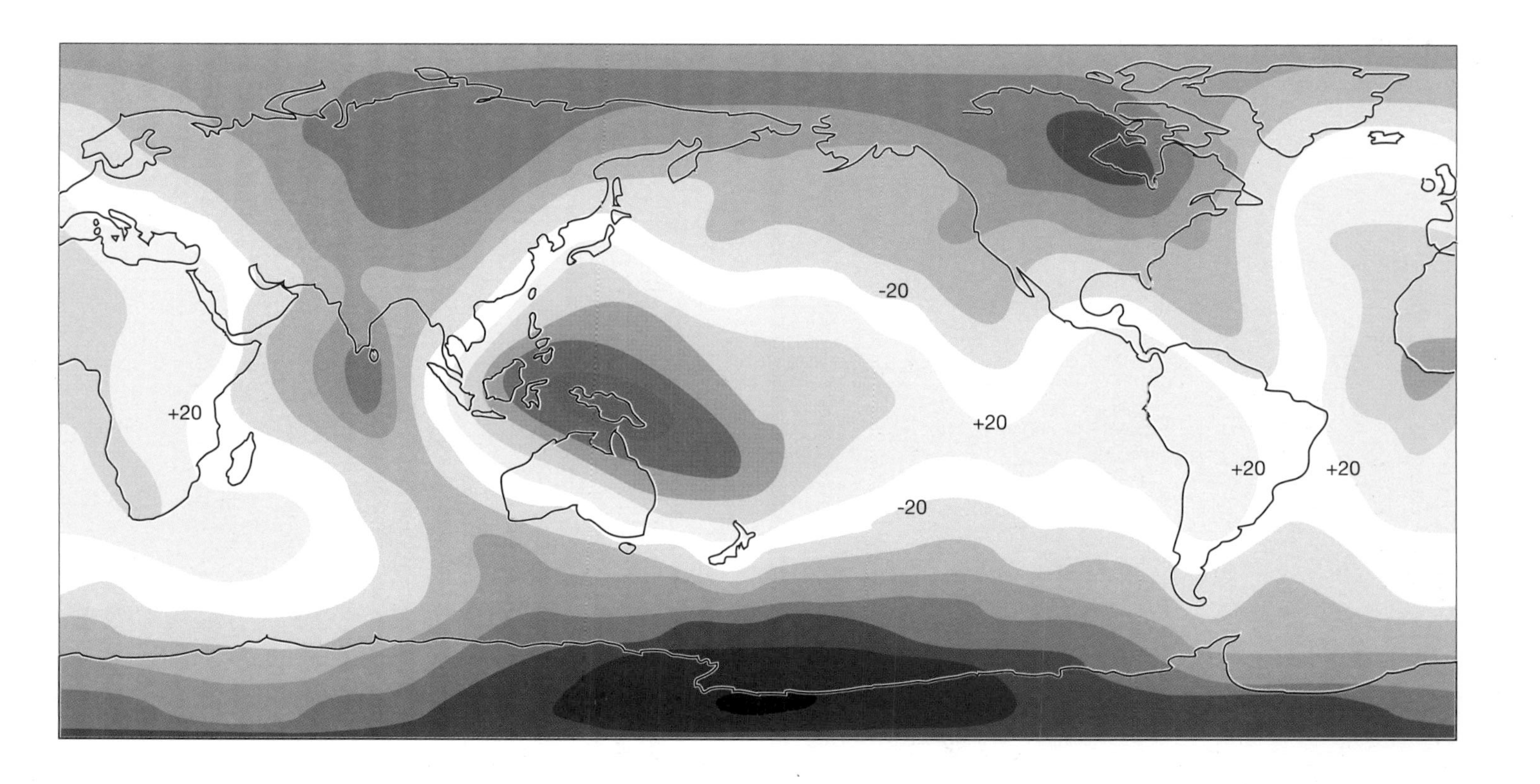

Figure 3.2
Very long wavelength anomalies of the geoid compared to the reference ellipsoid. The contours correspond to differences in elevation of the ocean surface in 20 m steps. The zero contour is located in the *white band.* Note the remarkable difference of 160 m between the depression in the Indian Ocean and the bulge in the western Pacific. (After C. G. Chase 1979, Nature 282, 466-468)

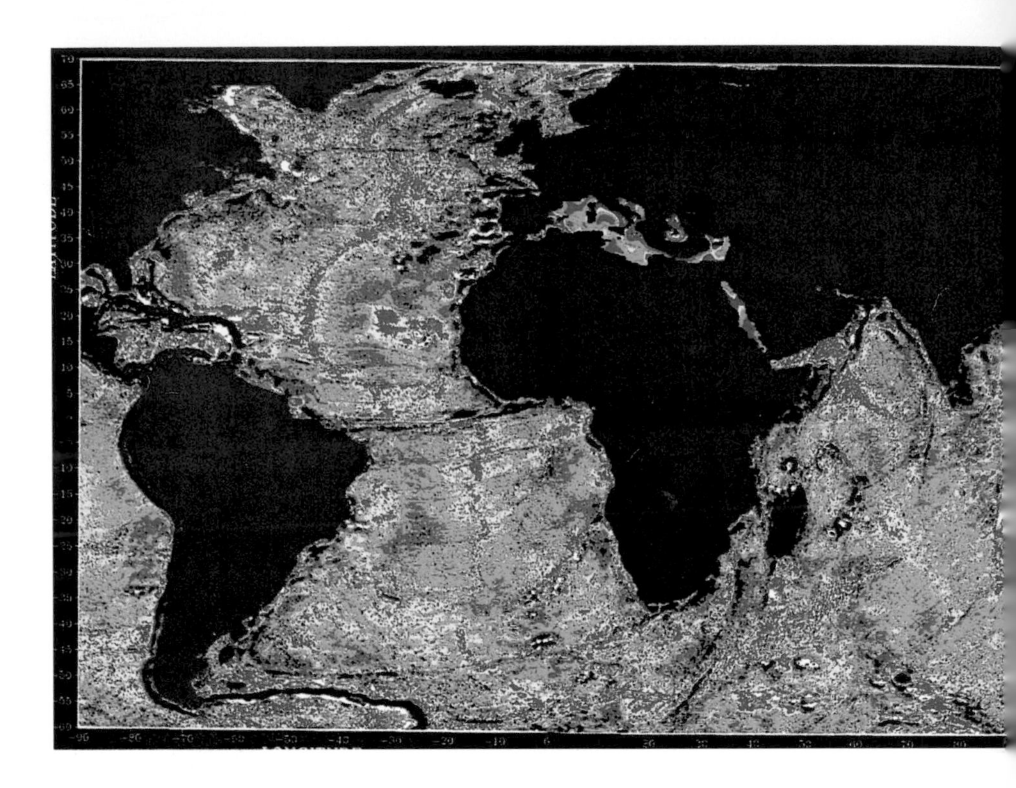

Figure 3.3
Is this a map of the ocean floor? No, of its surface! The highs and lows of the ocean surface as derived from satellite measurements reproduce with astounding fidelity the relief of the ocean floor. The map was constructed by selective filtering out the longer wavelength anomalies of the geoid in relation to the reference body. The ridges, and especially the East Pacific Rise, do not show clearly as the lithosphere here is very young and consequently very thin and soft. It cannot support a larger load and consequently deforms and is close to gravitational equilibrium. In this case, the ocean surface is not disturbed. The relief becomes much more prominent over old lithosphere which can sustain pronounced differences in load. Particularly notable is the deep groove over zones along which lithosphere is subducted. The Hawaii Emperor chain of volcanoes discussed in connection with hotspots is also clearly recognizable. (Altimetric data from Seasat satellite; map supplied by A. Cazenave)

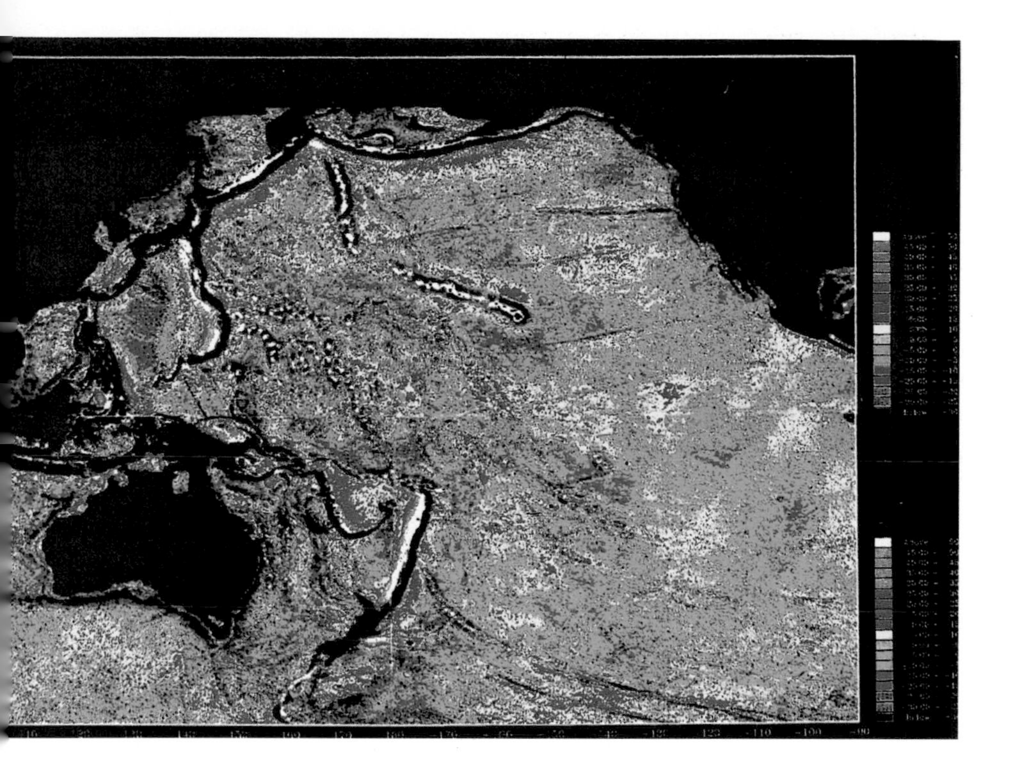

tances in the range of 100 km. To understand these highs and lows, we must know that this evidently mobile surface is highly sensitive to local, internal mass variations. Thus, when a submarine volcano is formed, its mass will locally load the lithosphere, thereby producing a local distortion of the gravity field, i. e. a high on the equipotential gravity surface. The ocean surface, which itself also represents an equipotential surface, will then also tend to form a high.

The geoid anomalies in the range of metres to tens of metres and of relatively short wavelength, a few hundred kilometres, correspond to high on the ocean surface above a positive submarine relief and to lows above a submarine trench. Mapping by satellite the highs and lows of the ocean surface is thus equivalent to mapping the sea floor itself (Fig. 3.3).

The anomalies of long wavelengths in Fig. 3.2 are related to internal factors, the main one being convection during which hot currents lead to negative mass anomalies and cold currents to positive ones, together with the ensuing dynamic pressures. The analysis of these anomalies from their effects on the geoid is rather difficult. To analyze convection from satellite altimetry, the data spectrum has to be filtered mathematically (Fourier analysis) in order to retain only the longer wavelengths.

The Belt of Rifts and Ridges

In the map of the ocean floors (Fig. 1.1) the ridges appear as the crest of this long backbone which run through the oceans, and occasionally divide them almost symmetrically as in the Atlantic. The crest is sometimes marked by an axial valley, the rift, which is particularly well developed in the Atlantic, as illustrated by the topographic section at right angles across the ridge (Fig. 3.4a). The term rift, here applied to the axial valley of certain oceanic ridges, is borrowed from continental structures like the East African Rift, the topography of which is comparable and which also represents extension features, as we shall see. The East Pacific Rise does not possess a central rift and its summit zone displays a much less rugged relief than the Mid-Atlantic Ridge (Fig. 3.4b, c), illustrating a profound difference between the two ridges.

Let us follow the ridges in Fig. 3.5. We observe that they constitute a system which encircles virtually the entire globe. Let us start from Iceland, an island built on the Reykjanes ridge as the result of an exceptional rate of volcanic activity related to the existence of a hotspot. To the northeast of Iceland, the ridge continues towards the pole, terminating north of Siberia. To the south, our ridge snakes its way at equal distance from either coastline towards the South Atlantic, where it turns east and splits into two at a good distance from the Cape of Good Hope. Following the eastern branch to the east of the island Réunion, we encounter a major division. The North Indian branch continues into the Gulf of Aden, where another division takes place. To the north, the oceanic rift of the Red Sea ends in the cul-de-sac of the Gulf of Suez, and to the south another branch continues into the East African Rift system. We can thus conclude that there is continuity between oceanic ridges and continental rifts and that the two structures belong to the same system of continental and oceanic expansion which may be traced over 75000 km. Let us now return to the central Indian Ocean to follow the branch of the ridge which here veers off to the southwest. It passes between Australia and Antarctica into the South Pacific, to turn north eventually, gradually approaching the American coast. It cuts it in the Gulf of California and returns to the ocean of British Columbia. Following

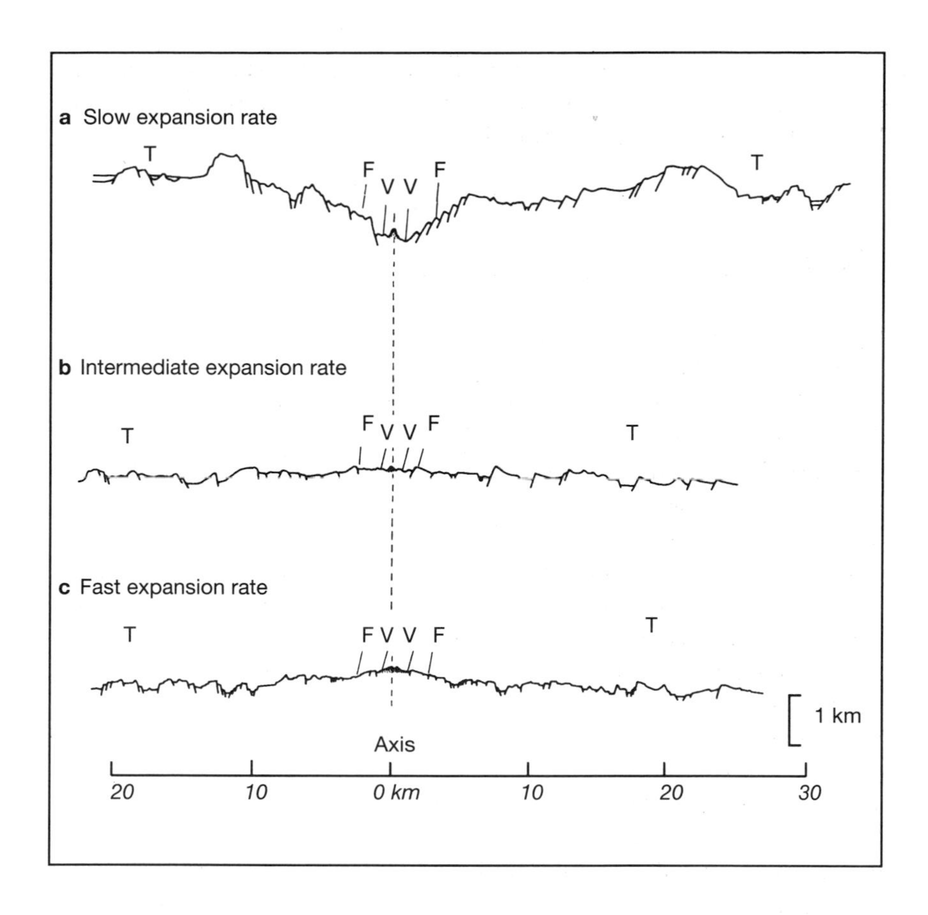

Figure 3.4 a–c
Topographic profiles across ridges. **a** Mid-Atlantic Ridge (1–2 cm/a). **b** East Pacific Rise at 21 °N (5–9 cm/a). **c** Equatorial zone of East Pacific Rise (over 9 cm/a). Note the narrow width of the zones of active volcanism (*V*, approx. 1 km), opening of fissures (*F*, < 5 km), and activity of faults (*T*, 20–40 km). (K. C. McDonald 1982, Ann. Rev. Earth Planet.Sci., 10, 155-190). Reproduced by permission of Annual Review of Earth and Planetary Sciences, vol. 10, © 1982 Annual Reviews Inc.

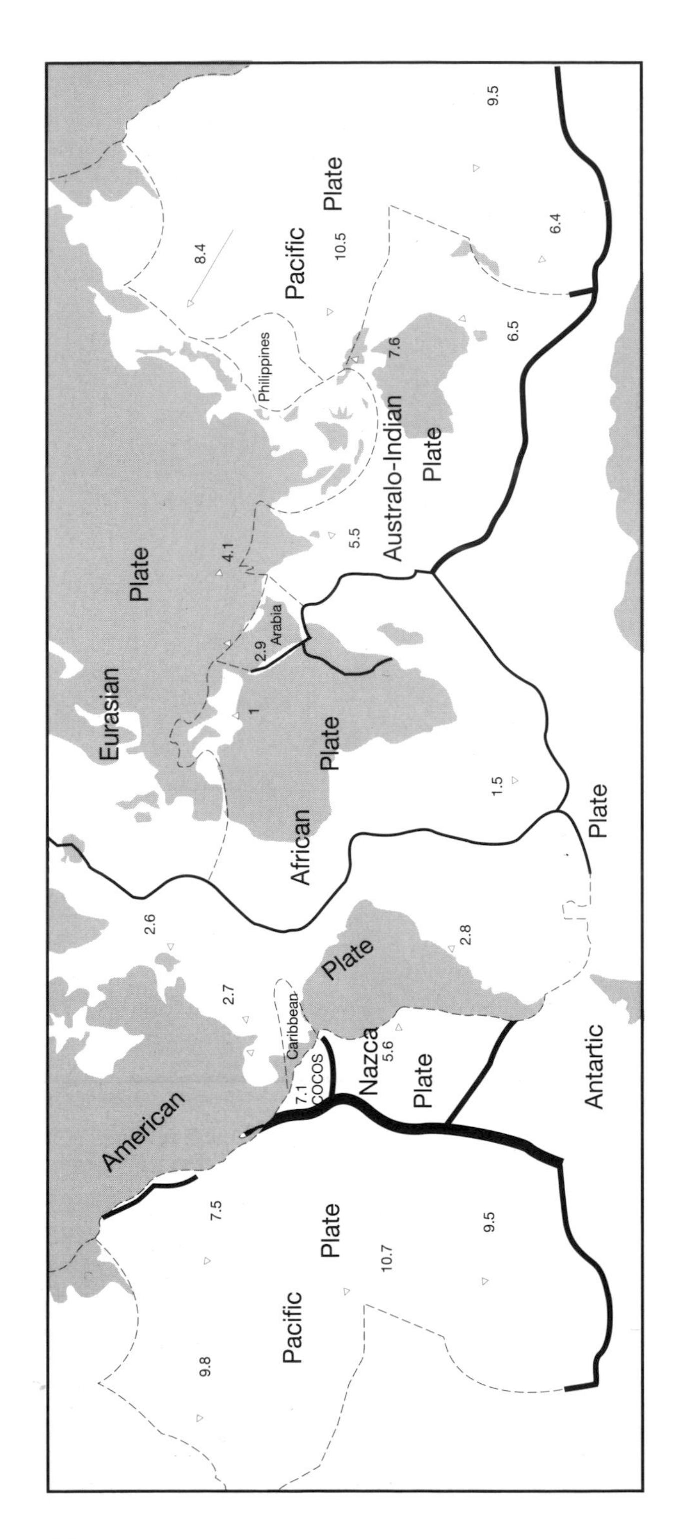

Eurasian Plate
Pacific Plate
Australo-Indian Plate
African Plate
American Plate
Pacific Plate
Nazca Plate
Antartic Plate
Philippines
Arabia
Caribbean
COCOS
8.4
10.5
9.5
6.4
6.5
7.6
5.5
4.1
2.9
1
1.5
2.6
2.7
2.8
5.6
7.1
7.5
10.7
9.5
9.8

◀ **Figure 3.5**
The major plates and the location of ridges and rifts. The thickness of the continuous line decreases with decreasing spreading rate from fast to intermediate and slow-spreading ridges and rifts. *Dashed lines* show the boundaries of the transform and compressive plate margins (subduction and collision chains). The *numbers next to the arrows* give the spreading rates of the various plates in cm/a. (K. C. McDonald 1982, Ann. Rev. Earth Planet. Sci., 10, 155–190 and J. B. Minster and T. H. Jordan 1978, J. Geophys. Res., 83, 5331–5354)

bone of ridges and rifts, we have passed over certain smaller branches like the Macquarie or Cocos ridges. All these are referred to as mid-oceanic ridges. This is an unfortunate term as it implies that they always result from the separation of two continents keeping always the same distance from either of the two resulting shorelines. This is clearly the case for the Mid-Atlantic Ridge but by no means so for the East Pacific Rise, which is bordered by a subduction zone along the west coast of South America. These ridges are in contrast to other ridges which open up in basins behind island arcs, the so-called back-arc basins, like the Fiji basin, which are numerous in the western Pacific Ocean.

Fracture Zones

It is appropriate now to dwell on certain incisions which cut and displace the ridges. As they are mostly oriented more or less at right angles to the ridges, they are referred to as transverse faults (Fig. 3.6). Other terms used are fracture zones or transform faults. Although the term "transform" should be reserved for those faults which really "transform" the opening movement of the ridges into another movement, the term has been established by now in the literature in a less restrictive sense. In a truly "transform" case the fault acts as a relay, for example the Azores-Gibraltar transform fault of the central Atlantic, which "transforms" the opening of the ridge near the Azores into a convergent and shear movement between Europe and Africa near Gibraltar (Fig. 3.5).

The most spectacular fracture zones of the central Atlantic are the Vema and the Romanche. The floor of the fracture is a groove some 5000 m deep with a rather flat bottom which is covered by slumps derived from the walls of the enclosing escarpments. These about 3-km-deep incisions into oceanic crust offer some prime geological

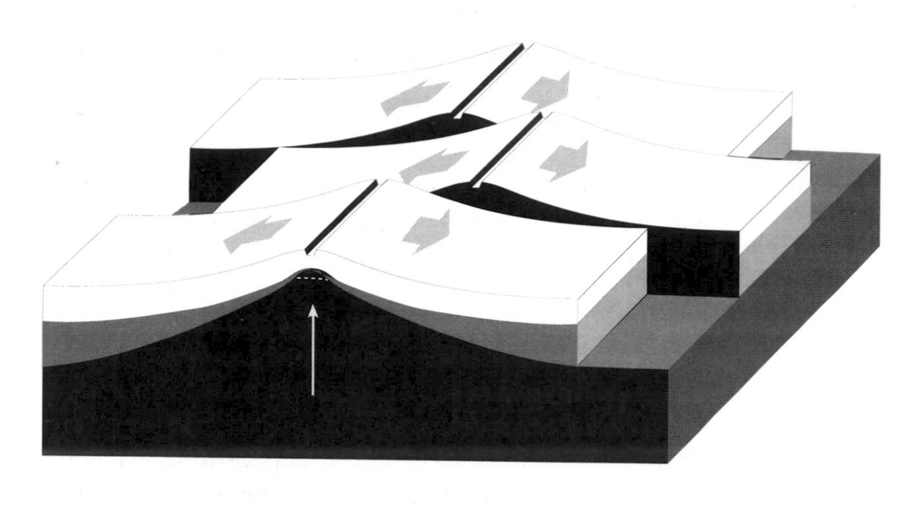

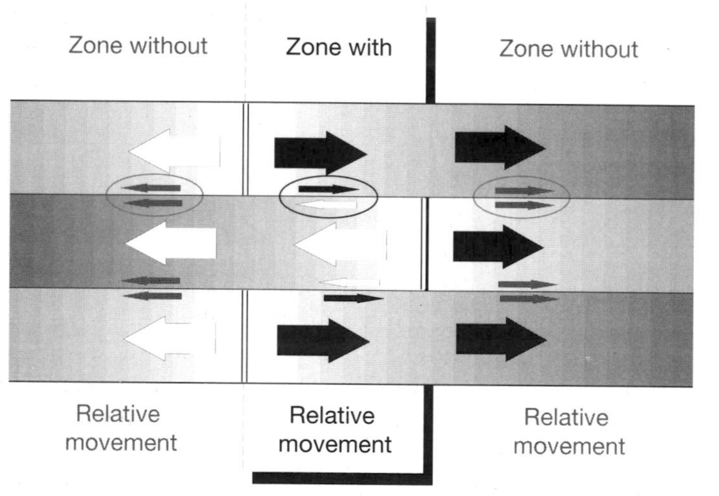

Figure 3.6
Ridges and transform faults. The block diagram illustrates the drawer-like functioning of a transform fault. The "drawer" consists of the lithosphere created along the ridge (crust shown in *white*, mantle in *light grey*) which slides on a floor of asthenosphere (*dark grey*). The plan view (below) shows that the faults are active only between the segments of the ridge itself because of the opposing sense of movement of the freshly created plates. Beyond this zone, the fault is inactive as the plates theoretically slide at the same velocity. We talk about the inactive zone or fracture zone and the active zone or transform fault. Note also that the sense of true relative movement in the active zone is opposite to the sense of offset of the ridge by the fault

sections. Here in 1988 the new French submersible NAUTILE (Fig. 3.7) during a dive to the Vema fracture zone, recognized and sampled a virtually complete section through oceanic crust which we shall discuss later (Chap. 4).

All these faults, large or small, displace segments of the ridge against each other. This displacement implies sliding movements on either side of the vertical fault planes (Fig. 3.6) which extend over several tens of kilometres in the case of the small- and medium-sized faults, and may reach several hundred kilometres in the case of large faults of the central Atlantic. When we analyze these movements in more detail keeping in mind the opening of the ridges, we note that the observed displacement represents only a part, and probably only a small part, of the true amount of shearing as explained in Fig. 3.6. Thus, between the two parts of the ridge in which the opening movements are in opposite directions, the transform fault sustains formidable shearing movements which cut through the entire oceanic lithosphere before grading into soft deformation in the underlying asthenosphere. These shearing movements make themselves felt also by numerous earthquakes resulting from the rupturing of rocks along the transform fault.

The fracture zones which cut vertically right through the entire lithosphere constitute prime pathways for its displacement. The portion of a ridge between two fracture zones thus opens like a drawer (Fig. 3.6). The drawer is a prism of lithosphere, its bottom being made up by the surface along which lithosphere slides on asthenosphere. The sides which guide the opening are represented by the two fracture zones. In this respect, the fracture zones show that the direction of displacement of the plates is necessarily parallel to the surface trend of these fractures as long as we disregard minor components of convergence or divergence between the two sides of the fracture.

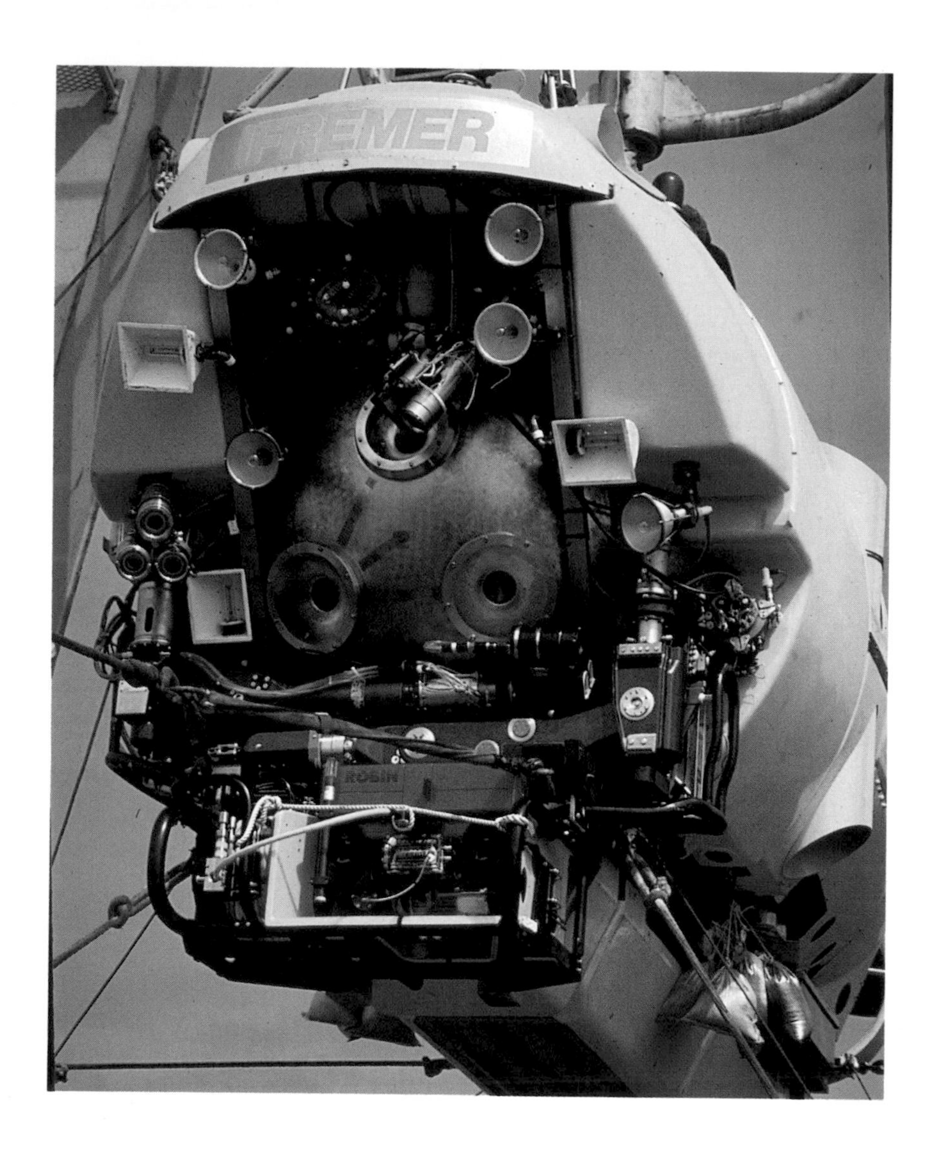

Figure 3.7
The NAUTILE, a French submersible capable of diving with three passengers down to a depth of 6000 m. (Photo courtesy of IFREMER, Nautile)

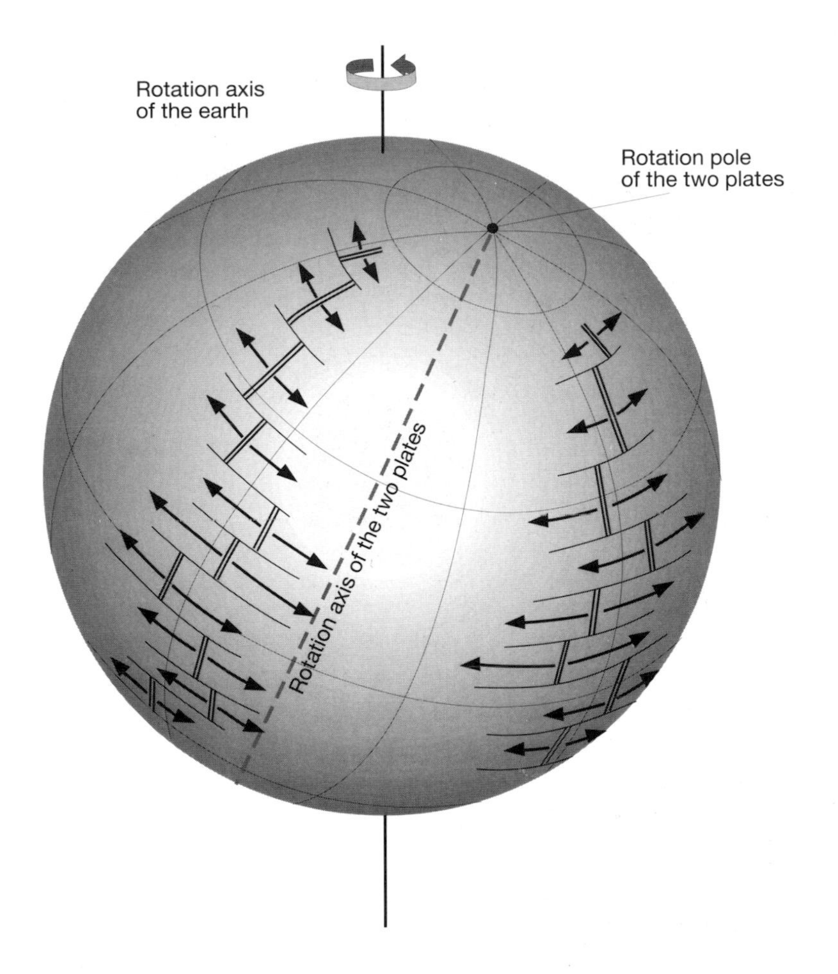

Figue 3.8
Movement of two plates expressed as a rotation around a so-called Eulerian pole. The movement increases from zero at the pole to its maximum at the equator. The transform faults represent parallels of the Eulerian sphere

Kinematics on a Sphere

The Swiss mathematician Euler showed that the relative movement between two rigid plates on a sphere will be defined by a rotation related to an axis, the Eulerian axis, and an angular velocity which is zero near the Eulerian pole where the axis pierces the sphere, and highest along the Eulerian equator (Fig. 3.8). The transform faults are then in principle the parallels of the Eulerian sphere.

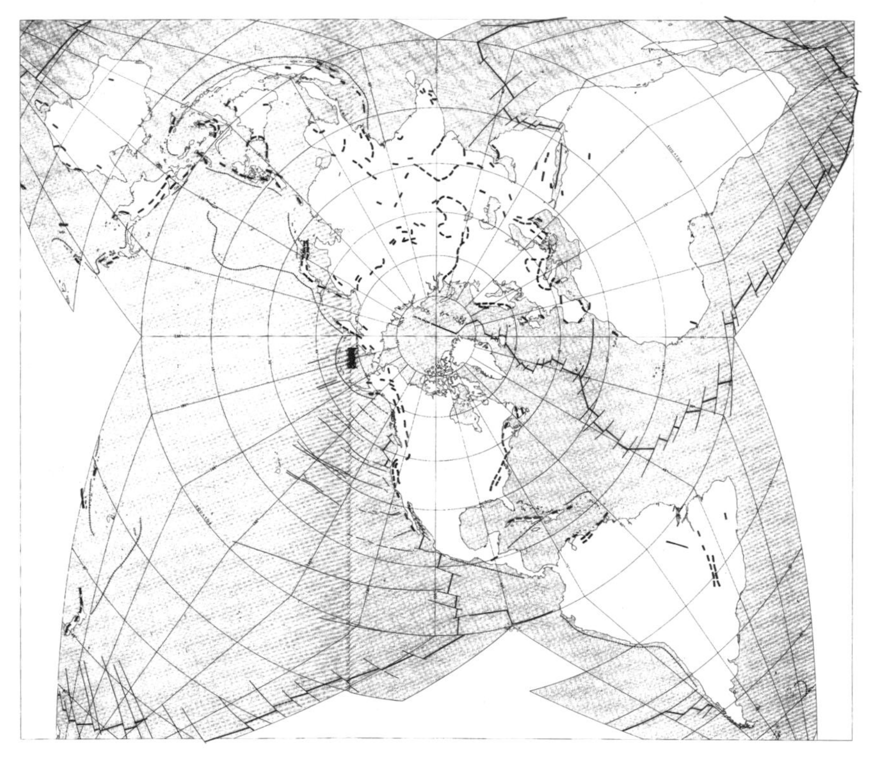

Figure 3.9
A projection of the Earth's surface centred at the North Pole, showing that the transform faults of the Mid-Atlantic Ridge and the East Pacific Rise are approximately parallel to the lines of latitude. This suggests that the poles of opening of the ridges are situated close to the Earth's axis. The *thick dashed lines* indicate the major ophiolite belts. (W. P. Irwin and R. G. Coleman 1974, U. S. Geological Survey)

The projection chosen in Fig. 3.9 has the advantage of illustrating the more or less N-S trace of the two main ridges, viz the Atlantic Ridge and the East Pacific Rise. In this case the transform faults represent more or less exactly parallels of the Earth's globe. This general arrangement results from the fact that the opening axes of both ridges are close to the rotational axis of the Earth (Fig. 3.9). It follows from this arrangement that the opening velocity increases in a regular manner from zero near the rotational axes towards the maximum along the Eulerian equator, which coincides more or less with the Earth's equator. The fracture zones represent here the means for accommodating the differences in opening velocity between the various portions of the ridges.

Slow- and Fast-Spreading Ridges

The analysis of the magnetic anomalies has told us that the opening velocity of ridges, also called spreading rate, varies mainly from 1 cm/a or less in an oceanic rift like the Red Sea to almost 20 cm/a on the equatorial part of the East Pacific Rise (Fig. 3.5).

Detailed exploration of the ridges started with Franco-American dives of the FAMOUS campaign. Conducted in 1973–74 in the Central Atlantic, it involved the submersible ARCHIMEDE of the French Navy alongside two new submersibles, the American ALVIN and the French CYANA. This remarkable campaign furnished the first clear images of the topography of the axial valley of the ridge and its intersection by a fracture zone.

The concept of the oceanic ridges was dominated by this image until an entirely new world was discovered by the systematic exploration of the East Pacific Rise, a campaign that represented the scientific highlight of the 1980s. Initiated in 1978 as a new project with the submersibles ALVIN and CYANA, exploration of this ridge started in a spectacular fashion with the discovery of oases of life crouched around columns of hot black water, the black smokers, rising above the floor of the ridge (Chap. 4). These campaigns were followed by remarkable marine geophysical operations, mainly under the auspices of American institutes. Intended to probe the deep structure of the ridge by indirect means such as seismics, the operations revealed, among other results, the presence of magma chambers below the ridge axis.

Segmentation of the Ridges

Systematic cartography of the East Pacific Rise since 1982 has shown a type of longitudinal subdivision of the ridge, also referred to as segmentation, which differs from the one known in the Atlantic. Whereas the Mid-Atlantic Ridge is cut every 50 km or so by a transform fault, along the course of the East Pacific Rise there is a remarkable continuity of the ridge between transform faults which are separated from each other by variable distances but in the range of several hundred kilometres (Fig. 3.10). The stretch between two faults, nevertheless, is interrupted, again at variable intervals of around 80 km, by strange hook-shaped structures, the Overlapping Spreading Centres of the American geophysicist Ken MacDonald. Between these zones, the ridge is remarkably regular in shape and direction provided we neglect local faults with a few hundred metres

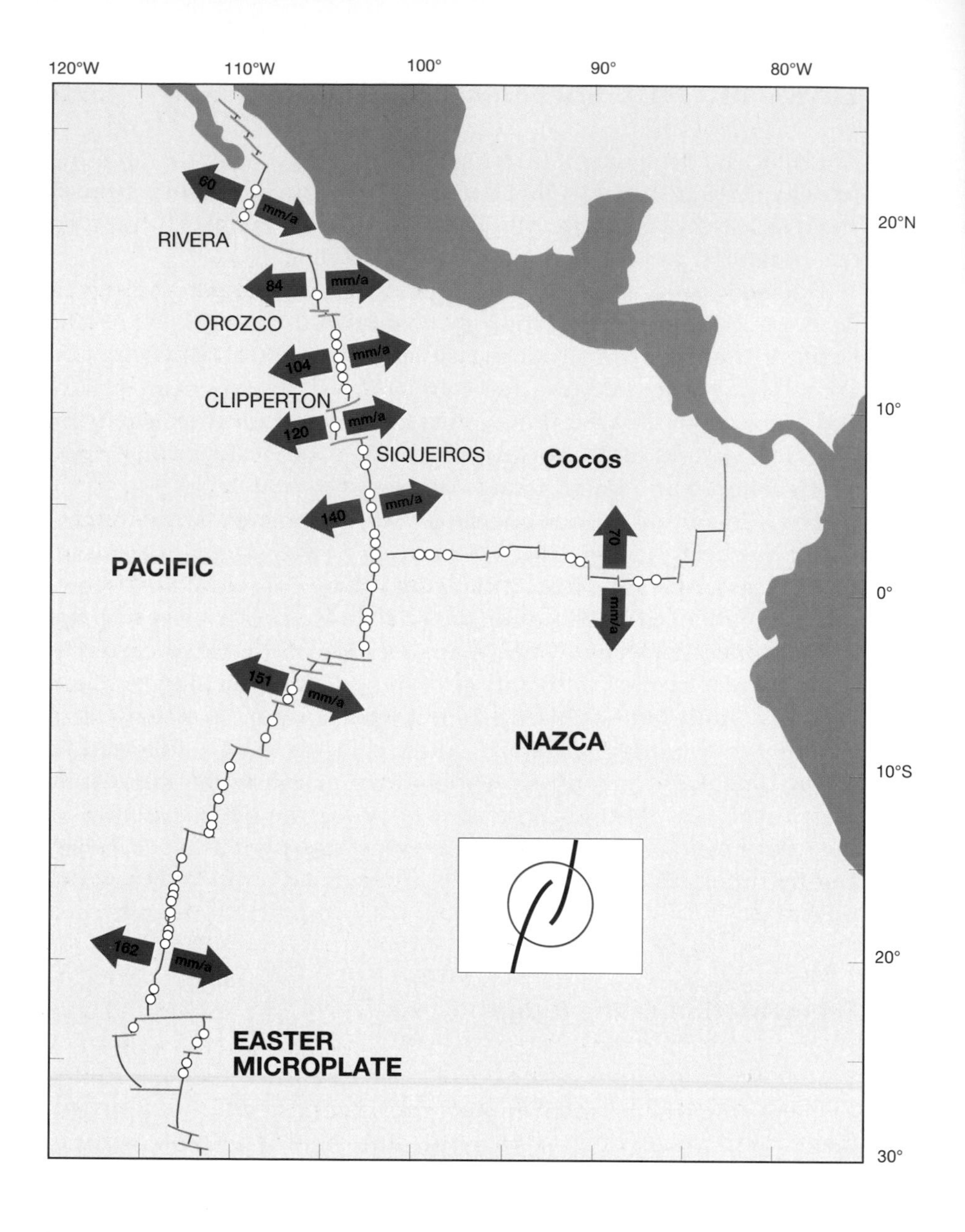

Figure 3.10
The segmentation of the fast-spreading ridges is determined by fracture zones and overlapping spreading centres. (K. C. MacDonald et al. 1986, J. Geophys. Res., 91, 10501–10511)

of displacement whose longitudinal periodicity is of the order of 10 km and seamounts which are volcanic peaks aligned on either side of the ridge (Fig. 3.1).

The segmentation of the East Pacific Rise is also illustrated by depth variations. A topographic profile parallel to the ridge (Fig. 3.11) is festoon-shaped with relief culmination at around 2600 m depth corresponding to the centre of the segments and topographic depressions to the two extremities. At a scale of several hundred kilometres, we observe another segmentation underlined by the deepest depressions plunging to below 3000 m and with overall differences of elevation in the range of 400 m. At a distance of 50–100 km, the highs and lows of such a segment oscillate by only about 100 m in elevation. The segmentation of the Atlantic ridge, mainly defined by the spacing between fracture zones, is characterized by differences in elevation from 200 to 2000 m between the culminations of the axis and the depressions marked by fracture zones. The mean differences in elevation here are thus much higher than on the East Pacific Rise. The depressions are also more pronounced when the horizontal displacement along the fracture is high. They reach up to 5 km over the South Indian ridge and offer remarkable sections through the crust.

In contrast to this segmentation which at the exaggeration usually chosen for such longitudinal profiles appears like a saw-tooth pattern (Figs. 3.11 and 3.12) we may discern a less abrupt depth variation on a scale of 1000 km or more. This becomes especially evident on the Atlantic Ridge (Fig. 3.12). Starting from a position above sea level on Iceland, a mean depth of 3500 m is reached some 1500 km to the south. The crest then rises to -2000 m near the Azores to fall again to -3500 m some 1000 km further to the south. The flexures of the sea floor on this scale may be related to mantle plumes which bring magma and heat from the deeper mantle.

Data now available suggest that, compared to the Atlantic Ridge, the East Pacific Rise is structurally rather simple, reflecting an equally simpler deep structure and mode of functioning. Because of this, we shall revert to the East-Pacific Rise as a reference when discussing the deep structures and functioning of a ridge in Chapter 9. However, when studying the nature and chemical composition of oceanic crust, dissections of several kilometres along fracture zones in slow ridges prove more useful than the modest depressions only a few hundreds of metres deep of the fast ridges. We shall then describe the oceanic rocks from a section on the Atlantic Ridge.

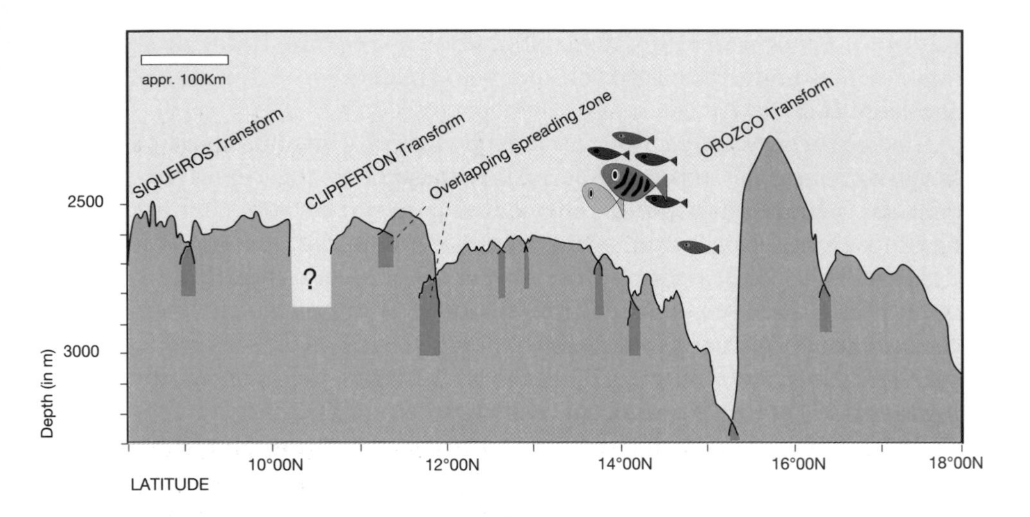

Figure 3.11
Section along the East Pacific Rise (vertical scale greatly exaggerated) illustrating the different scales of segmentation. The major discontinuities are fracture zones like the one at 10 °N (Clipperton Fault shown in Fig. 3.1). On a smaller scale, the overlapping spreading centres are shown in darker colour. (After K. C. MacDonald et al. 1988, Nature 335, 217–225)

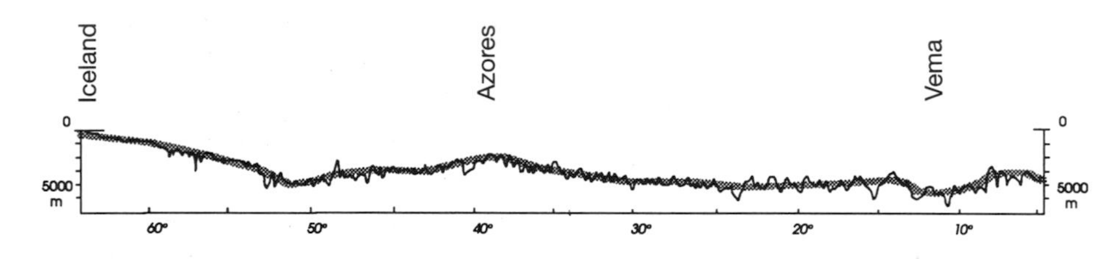

Figure 3.12
Profile along the Mid-Atlantic Ridge from Iceland towards the equator. On the saw-tooth-like segmentation caused by fracture zones, an undulation on the scale of thousands of kilometres is superimposed, which is related to the activity of hot-spots. (P. Gente 1987, Doctoral thesis Univ. Brest, 371 p.)

Oceanic and Continental Rifts

Our tour of the world of expansion (Fig. 3.5) has shown that oceanic rifts like the Red Sea or the Gulf of Mexico and continental rifts like the East African Rift system are extensions of the oceanic ridges and clearly belong to the same global system. In contrast to this, the formation of other continental rifts like the Siberian Baikal and, in Europe, the Upper Rhine Graben and the Limagnes of the French Massif Central could have been favoured by a fracturing of the lithosphere under the influence of a plate collision.

The morphology of the rifts is similar to that of slow-spreading ridges. Continental rifts are characterized initially by a relief of 1500–2000 m over a width of several hundred kilometres. Following lithospheric extension, the keystone of this arch founders (Fig. 3.13) to form an axial valley several tens of kilometres wide. The Vosges Mountains and the Black Forest thus frame in a symmetrical fashion the collapse of the Upper Rhine Graben.

The regional relief or swell is provoked by lithospheric thinning and the rise of asthenosphere. Actually, in our balance of mantle weights (Fig. 1.10), the replacement of about 100 km of lithosphere by a thinner lithosphere results in a weight decrease in the corresponding column. As the underlying asthenosphere is hotter than the lithosphere and thus also lighter, the relief will increase (in the model we replace air by denser rocks) and the equilibrium between the scales of the balance is reestablished. We shall see in Chapter 6 that the ascent of asthenosphere to depths of less than 75 km will lead to partial melting and to the rise of basaltic liquid. This will account for volcanism, which represents another manifestation of the activity in rifts.

The stretching of continental rifts with the axial valley as the most obvious expression is also marked by thinning of the crust by 5–10 km from about 30 km to 20–25 km. Oceanic rifts are derived from continental rifts which as a result of continuing extension are split apart, facilitating the formation of oceanic lithosphere along the rift axis. Prior to becoming completely separated, the shoulders of the continental rift experience considerable stretching and the 30-km thickness of the continental crust may be reduced to 5–10 km (Fig. 3.14). This stretching may be measured on surface by the displacements on either side of faults which become inclined at depth, the so-called listric faults (Fig. 3.15).

An oceanic rift is destined to develop by progressive opening into a mid-oceanic ridge like that in the Atlantic. The marginal continental zones bordering an oceanic rift or a mid-oceanic ridge are called

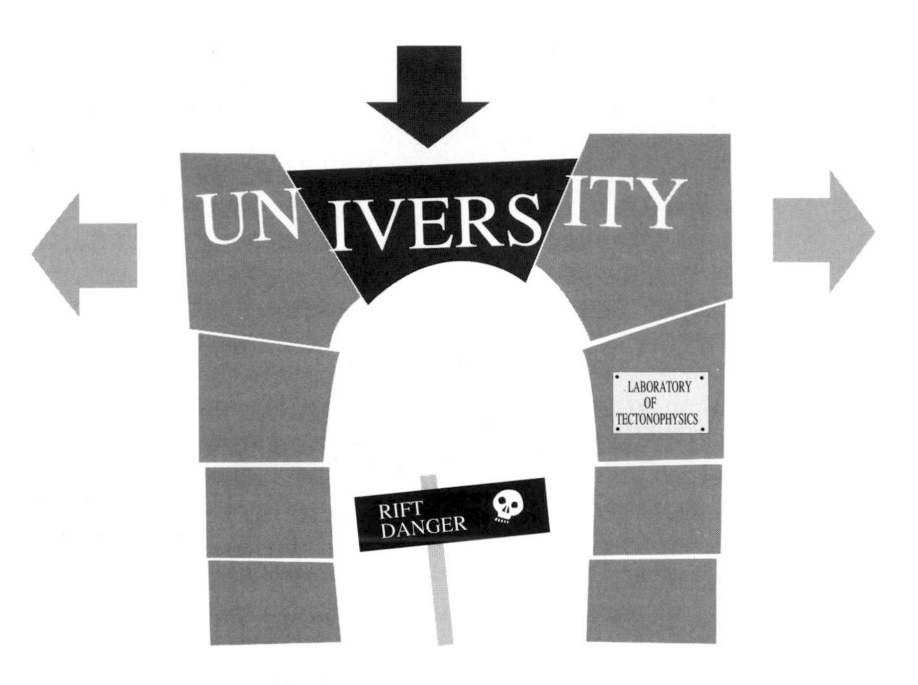

Figure 3.13
Architectural model showing the sinking of the keystone of a vault resulting from a slight displacement of the walls, illustrating the sinking of a rift between two shoulders subjected to tectonic extension. The planes along which the keystone slides down are called normal faults

"passive margins" in contrast to the active margins which border subduction zones. Passive margins, stretched and sinking because of thermal subsidence, are important traps for sediments and sometimes also hydrocarbons. Prospecting for and production of hydrocarbons takes place off-shore along such passive margins.

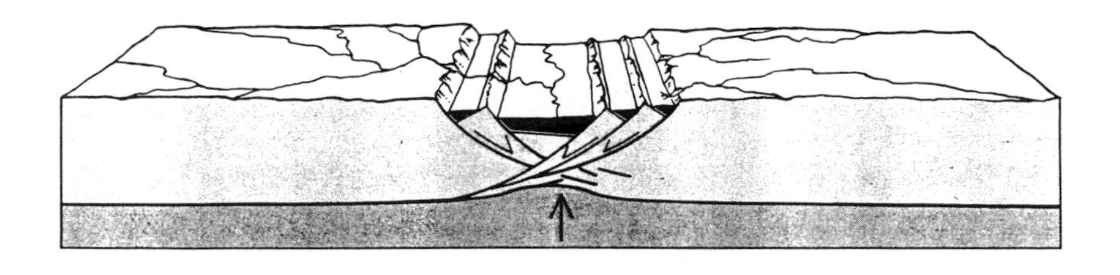

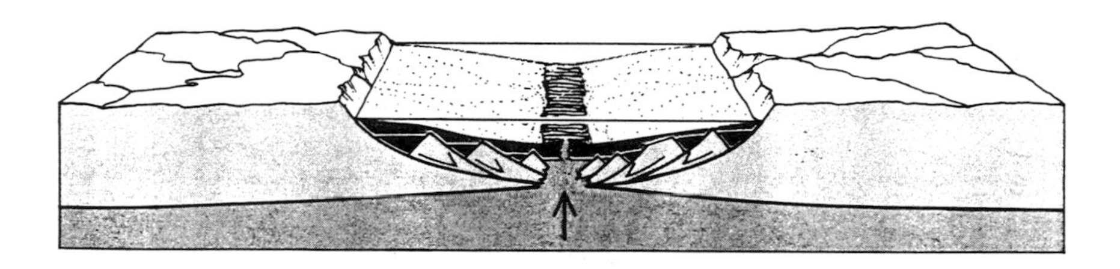

Figure 3.14
Development of an ocean from a continental rift. During the initial stage (*above*), the crust becomes thinner. At the second stage, like in an oceanic rift of the Red Sea-type, the continental crust recedes and oceanic crust forms. In the final stage of a passive margin, the axis of the ridge has moved far away from the sides of the rift. In this stage, the crust has cooled, contracted and as a result subsided (thermal subsidence), thereby favouring active sedimentation. (C. Burchfiel 1983, Sci. Amer., 249, 86–98)

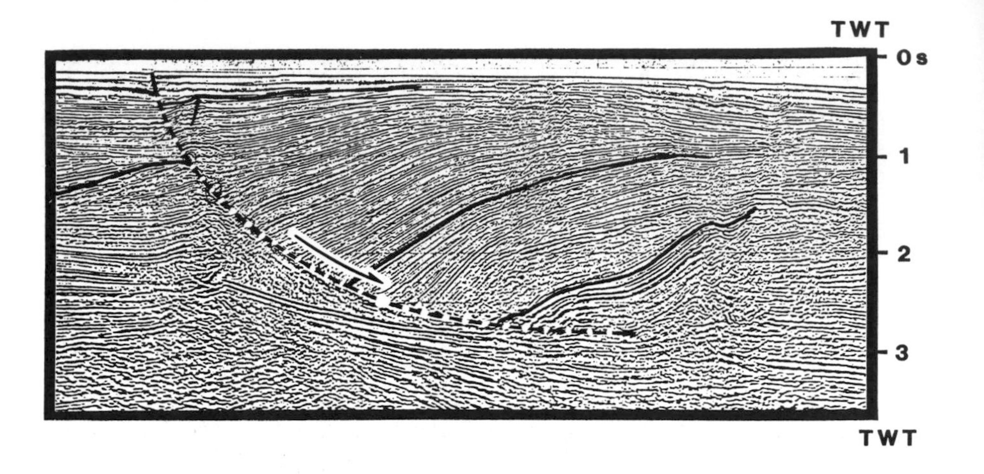

Figure 3.15
Seismic profile (vertical seismic reflection) in the sedimentary fill of a rift illustrating the extension resulting from slumping along a curved, so-called listric fault (*dashed*). The curvature of the fault leads to a rotation of the displaced beds (*thick lines*). The vertical scale (in s) represents the time required by the seismic waves for the two-way trip. At common wave velocities in sediments, 3 s correspond to about 5–6 km. (After V. L. Faure and J. C. Chermette 1989, Bull. Soc. Géol. Fr., (8), V 3, 463 modified)

4 Submarine Exploration

Dredgings, drill holes, studies from submersibles, seismic soundings and geophysical modelling have revealed that the crust on average basaltic in composition is layered and separated from the mantle by the Moho. It forms very close to the ridge axis, which is like a long permanently open wound. Some 20 km from the axis the lithosphere is already scarred and passively moves farther away. Hydrothermal activity, which extinguishes the fire of the ridge, maintains chimneys belching hot black water, surrounded by oases of life without light.

The Oceanic Rocks

The most simple means for exploring the nature of the rocks constituting the floor of the ocean is dredging. A slowly travelling ship drags behind it a large cage open at the bottom, which closes when pulled up, to detach and hold rock fragments several tens of kilograms in weight. The method, unfortunately, is only a kind of blind poking, a shortcoming that is alleviated by the use of the grabs of submersibles. Finally, for the last 20 years, international drilling ship-borne programmes have furnished kilometres of samples in the form of borehole cores from oceanic rocks. The deepest of these holes, measuring over 2100 m, penetrated under 200 m of sediment cover some 1900 m of oceanic crust near the East Pacific Rise. This is still a rather modest value considering that the total thickness of the oceanic crust is around 6 km.

One of the best sections through the crust to date was recorded a few years ago by the submersible NAUTILE along one of the sides of the Vema fracture zone (Fig. 4.1). This section combines in an orderly sequence the principal oceanic rock types which were already long known from dredgings.

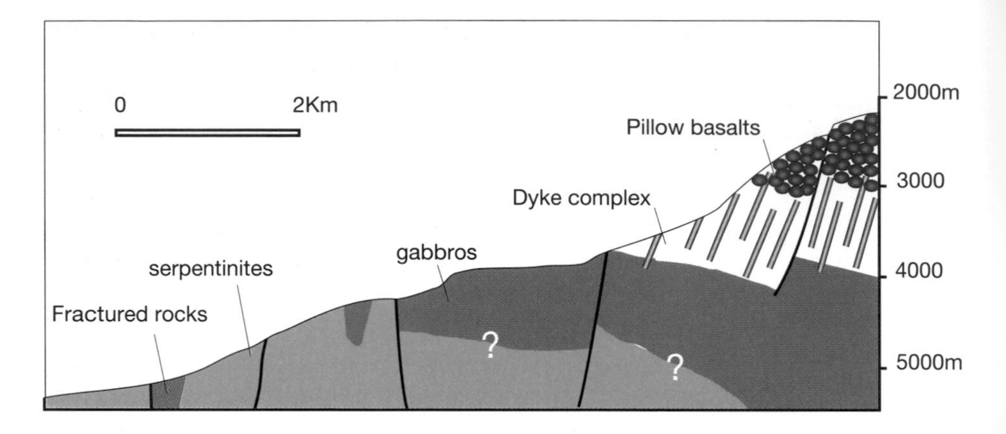

Figure 4.1
Geological section recorded by NAUTILE during a dive along the Vema fracture zone in the Atlantic. (After J. M. Auzende et al. 1989, Nature 337, 726–729)

The bottom of the fracture zone is made up by peridotites, rocks enriched in magnesium and depleted in alumina and calcium against the rocks of the crust and belonging to the underlying mantle. They have, however, been altered due to prolonged contact with sea water. Having taken up water to 1/6 of their weight, they grew in volume in this process and became altered to serpentinites, rocks that derived their name from their scaly nature and greenish colour which is reminiscent of the skin of a snake.

During its ascent along the wall of the fracture, the NAUTILE left the serpentinites at a depth of about 4500 m to encounter a completely different rock type, viz, the gabbros belonging to the oceanic crust. Between these gabbros and the underlying formations of the mantle there is an important boundary layer, the Moho. It separates the crust from the mantle below. As a simplification, we could say that the upper mantle and its constituting peridotites are dominated by one mineral, olivine. This is a magnesium silicate of olive green colour, sometimes forming gem-quality crystals of peridotite which were already treasured by the Pharaons. Olivine is also abundant in meteorites, suggesting that it is present also in the mantle of other planets. The crust is dominated by other minerals, the feldspars, which are aluminous silicates of a more complex and varied nature. In granites, the most common rocks of the continental crust, the feldspars are always enriched in the alkali elements sodium and potassium. In contrast to this, the feldspars in gabbros, the dominant rocks of the oceanic crust, are rich in calcium. The chemical composition of a gabbro is identical to that of basalt, the difference being the larger crystal size in the former which results from the slow cristallization of the original magma under the thick cover of basalts isolating it from the ocean water.

Continuing its ascent along the slopes and cliffs of the fracture, the NAUTILE eventually also encountered the basaltic cover. However, between this and the underlying gabbros, between -3000 m and -4000 m, the submersible encountered the so-called dyke complex made up almost exclusively of vertical basalt dykes intruding into each other. These dykes represent the pathways used by the rising basaltic magma to feed the overlying basalts. We shall examine these feeding mechanisms in Chapter 6. The last 800 m up to the top of the fracture are composed of basaltic lava flows and strange cushion-shaped lavas, the so-called pillow lavas. Their size ranges up to several metres and they derive their peculiar shape from the flow of lavas within a tube, the perimetre of which is chilled in contact with sea water (Fig. 4.2). the term tube lavas would thus actually be more appropriate. To make the section complete, we would have to include the cover of oceanic sediments accumulated on the basalts since the origin of this part of the crust on the Mid-Atlantic Ridge some 16 Ma ago.

Similar sections, although less complete, have been recorded in other parts of the Mid-Atlantic Ridge and on other slow-spreading ridges like the Gorringe Bank off the coast of Portugal or the South Indian Ridge. The outcrops of peridotites in the immediate vicinity of these ridges came as a real surprise. As submarine erosion, especially in such a young environment, would be negligible, we have to conclude that at least locally the ridge may be entirely lacking in oceanic crust. In contrast to this, along fast-spreading ridges, the basalt cover is rather continuous and it is difficult to find out what the basalts are actually covering. Because of this, a reference borehole was drilled near the East Pacific Rise in hard rocks of the oceanic crust at great technological expense to a depth of 1900 m into the ocean floor. The borehole managed to penetrate the basalt cover and ended in the dyke complex. In spite of several attempts over years, involving each time the retrieval of a borehole lost below a few kilometres of water somewhere in the Pacific Ocean, the horizon of the dyke complex has not yet been passed. However, the situation encountered so far illustrates the similarity between the upper part of the crust below slow- and fast-spreading ridges.

Figure 4.2
Pillow lavas in the Vema fracture zone on the central part of the Mid-Atlantic Ridge. (Photo IFREMER during the VEMANAUTE campaign, fracture zone VEMA, central Atlantic)

The Penetrating Eye of Geophysics

Geologists and Geophysicists: Square Your Accounts!

Whereas geology is a science based on the direct observation of terrestrial formations, geophysics employs indirect means, as the formations to be investigated do not crop out on the surface. Their objective is thus to understand the deeper, invisible zones under the guidance of geological field observations. The indirect means of geophysicists are frequently rather sophisticated instruments and the interpretation is generally based on the numerical data produced by these instruments. These instruments and the impressive calculations confer to geophysicists a certain prestige, or a "superstatus", which is sometimes observed jealously by geologists, who are armed only with their hammers and field boots. Geophysical arrogance and geological irritation have to be blamed for the lack of communication between the two branches, which slowed down the advance of the earth sciences for a considerable period. With the advent of plate tectonics, a dialogue became unavoidable, and a new, better relationship of cross-fertilization between the two communities was established. Measurements cover the field at an increasing rate and the geologist, who until now had been too occupied with descriptions, now extends his approach to the search for physical interpretative models, while the geophysicist learns to observe the Earth. Alongside the traditional geophysical fields such as gravimetry, magnetism and – most widely used – seismology, new disciplines like geophysics of solids have appeared, in which it becomes increasingly difficult to discern who, among solid-state physicists, crystallographers or mineralogists, is more a geophysicist or geologist. We welcome this pooling of abilities which now characterizes the larger research groups.

Structure of the Oceanic Crust

We know now that the oceanic crust, consisting of gabbros and basalts with a thickness of 6 km, rests along a certain surface, the Moho, on a mantle made up of peridotites. The presence of the Moho is well known to geophysicists because of its effects on the propagation of seismic waves. These waves result from natural earthquakes or are triggered by explosive charges. They travel within the Earth just like a wave on a water surface after a rock has been thrown in. When the wave triggered by such a rock encounters an obstacle like a wooden plank, it rebounds along a different path, and we then say that it is reflected. This also applies to the seismic waves within the Earth. Because of their physical properties related to their density, which is notably higher than that of the crustal rocks, the mantle rocks act as a screen against the seismic waves just like the plank on the water surface. The seismic wave coming from the surface may thus be reflected (or refracted) by the interface between crust and mantle. Taking into account the trajectory and the travel time of a reflected wave, we may calculate the position of such an interface. The interface between crust and mantle at a depth of 35 km below the continental crust was identified for the first time in 1910 by the Yugoslav geophysicist Mohorvicic, and in his honour it is usually referred to as the Moho. When repeated over the oceans, these measurements revealed that the Moho occurs at an average depth of 6 km below the oceanic crust, which thus attains less than fifth of the thickness of the continental crust.

With the advance of technology, we have since become able to identify two main interfaces within the oceanic crust, one on top of the gabbros and the other between the basalts and the sedimentary cover (Fig. 5.8). Recently, new refinements in seismic analysis have revealed under the East Pacific Rise the presence of chambers filled with basalt magma roofed by solidified walls (Fig. 7.12).

Ridge Dynamics

Under the spotlight of seismic waves, which here are indeed comparable to light waves, we have learned to penetrate the axial structure of the East Pacific Rise. Following the fundamental contribution to ridge expansion made by another geophysical technique, geomagnetics, and more especially by the magnetic anomalies presented above, we shall now demonstrate how other geophysical techniques like heat flux measurements, gravimetry or electrical conductivity measurements contribute to our understanding of the life and dynamics of ridges and to building models describing the structure and movements of crust and mantle below the ridges.

The heat flux is a measure of the amount of heat escaping per second over a unit surface area. On ridges, this flux is considerable as it removes the heat furnished by the ascent of basaltic magma. To the sides, the magma and then the rocks are cooling and the flux diminishes rapidly. Tied to the transfer of heat by conduction across a lithosphere which thickens away from the ridge on either side of it, the heat flux through lithosphere of increasing age decreases at a square root. This permitted us in Chapter 1 to estimate the thickness of the lithosphere and the depth of the oceans as a function of their age. This law is not fulfilled for the first few hundred kilometres from the ridge axis, where the flux is rather erratic. We attribute this unexpected behaviour to hydrothermal circulation through the crust, which is particularly intense near the ridge axis. Despite these disturbances, the distribution of thermal flux remains a record of the heat sources at depth and allows us to construct models of magma chambers and their enclosing environments under the ridges (Fig. 7.8).

These models may be improved if we take into account the gravity field over a ridge, as it is established that the magnitude of gravity at a given point is the function of the underlying masses. Variations in this magnitude are very weak but then may nevertheless be registered with exceptional precision by gravimeters. These variations from one point to the next result from differences in the distribution of masses further down below these points. Thus, a chamber filled with basaltic magma which is lighter than the solidified rocks along its walls will make itself felt by a mass deficit which, in turn, will give rise to a negative anomaly in the gravity field. It becomes clear that a model of a magma chamber based on the structural data of seismology will have to be compatible with thermal flux and gravimetric data.

In the field, matters become complicated, as the gravity signal is also influenced by dynamic processes. Let us assume that the mantle below a ridge exerts an upward pressure. The relief will then tend to rise to a height at which the weight of the corresponding column of rock counterbalances the mantle pressure. The additional weight thereby induced will create a positive anomaly. These new data have to be considered in our models. If actually the model of the magma chamber is already well constrained by data other than gravimetry, we may be in a position to extract this dynamic component from the gravity signal and thereby estimate the pressure exerted by the mantle below the crust.

We have so far focussed our attention on the crust below the ridge. Seismology tells us that in the mantle below this crust, the travel velocities of the waves and the quality of the respective signal decrease in a domain which is still poorly defined, but under a ridge, extends downward over a few ten kilometres. This is attributed to the presence of basaltic liquid in the mantle formations (Chap. 6). This interpretation is confirmed by the fact that the electrical conductivity of this zone is above that of the surrounding mantle, as the electrical conductivity is highly sensitive to the presence of liquid films which are better conductors than solid rocks. It must be stressed that the presence of basaltic liquid, the density of which is 5/6 that of peridotite, also produces a gravity signal which must be considered in our models. It is in principle fairly easy to estimate this effect, as the source is located at great depth and is of great extent compared to the magma chambers considered above. The wavelength of the signal will thus be notably longer and it could, if necessary, be eliminated by filtering.

To sum up, seismology, as the main geophysical technique constrains for us the framework of the ridges: the depth of the Moho, the main boundaries resulting from the crystallization of magma in the crust as it moves away from the ridge (Fig. 5.8) and now also the depth and shape of the roof of magma chambers below the ridge (Fig. 7.12). It tells us also that within the mantle below a ridge there is a zone extending over several tens of kilometres in height in which peridotites contain droplets and films of basaltic liquid like a sponge filled with water. It is obvious that this liquid feeds the overlying ridge.

Taking recourse to other geophysical techniques like measuring the amount of heat (or thermal flux) escaping from the oceanic crust, or the variations in the distribution of light and heavy masses (gravimetry) or, lastly, the conductivity for electrical currents under a ridge which is sensitive to conductive basaltic liquid, we are able to construct computer models of the functioning of ridges. We shall examine these after we have discussed the information on the same subject gleaned from the study of ophiolites.

Hydrothermal Activity of Ridges

April 20th, 1979

Red-blooded worms with their heads protruding from up to 2-m-long tubes, molluscs, also red-blooded and measuring some tens of centimetres, and swarms of crabs and shrimps in the night of the floor of the East Pacific Rise around a chimney belching out black water – this was the nightmarish panorama discovered by the small American submersible ALVIN on April 20th, 1979 within the framework of a French-American programme in which also the French submersible CYANA took part (Fig. 4.3). The show was not entirely unexpected as in 1977 ALVIN had already observed a bivalve colony on the Galapagos Ridge and in 1978 CYANA had discovered on the new site sulphurous structures surrounded by a dead colony. Actually, the existence of hydrothermal circuits cooling the ridges had been predicted many years before this date as an explanation for the thermal anomalies observed along ridges. After the discovery by CYANA, a relationship was rapidly established between the dead oases and the discharge points of hot hydrothermal liquid which would cause a notable rise in water temperature along the ridge. Based on a precise network of temperature measurements along the floor of the ridge, ALVIN was guided directly towards those hot zones where it was to discover the first living oasis. A large number of other thermal springs has been discovered since, along the East-Pacific Rise and then in the Atlantic, where, prior to this, only low-temperature seeps had been observed during the FAMOUS campaign in 1973. Recently, the submersible NAUTILE discovered such springs also along the ridge in the North Fiji Basin in a back-arc environment. Each of the ridges appears to possess its own biological signature, confirming the absence of exchange between ridges and independent biological evolution on each of them.

Figure 4.3
Submarine oasis around the chimney of a black smoker surrounded by a field of worms emerging from their tubes. (Photo IFREMER during the BIOCYATHERM campaign on the East Pacific Rise)

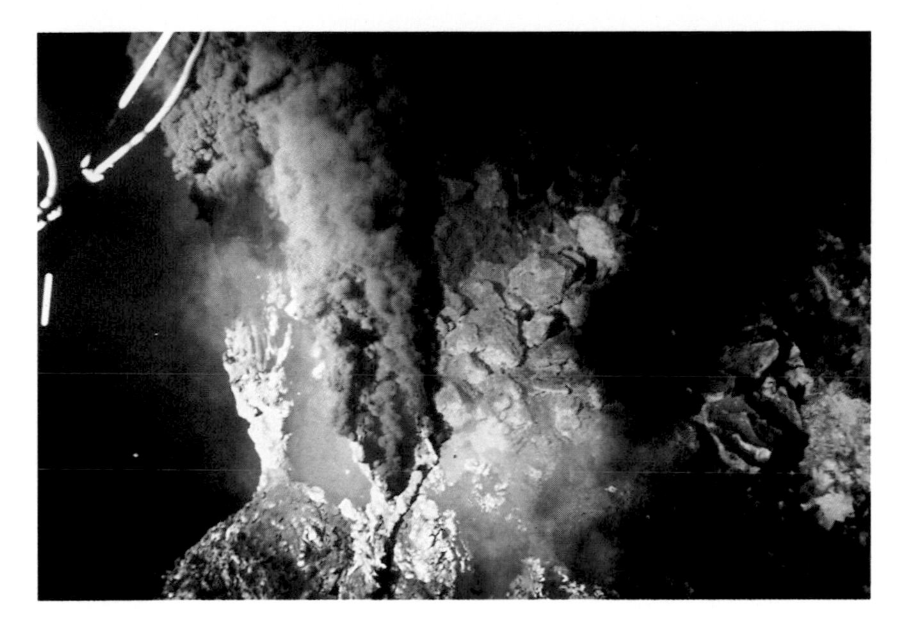

Figure 4.4
A black smoker. (Photo IFREMER during STARMER I campaign in the North Fiji Basin)

The Black Smokers

The hot springs on the East Pacific Rise are also referred to as black smokers, belching out waters of about 350 °C which obtain their colour from suspended metal sulphide particles (Fig. 4.4). The deposition of the sulphides around the vent progressively builds up a chimney which may reach several metres in height. The dissolved minerals and the heat given off by the black smoker favour the prolific growth of bacteria which, in turn, represent nourishment for the surrounding colonies.

The black smokers are the spectacular expression of an extensive hydrothermal plumbing system (Fig. 4.5). Tectonic dilatation experienced by the ridge over the first few kilometres on either side of its axis opens up fissures along which sea water percolates downwards. It easily penetrates the crust down to 2–3 km, i. e. to the very base of the dyke complex above the underlying gabbros, availing itself on its descent of the contacts between the vertical dykes. Farther down, penetration of water becomes more difficult due to the massive nature of the gabbros. Heated to 400–450 °C, the water circulates

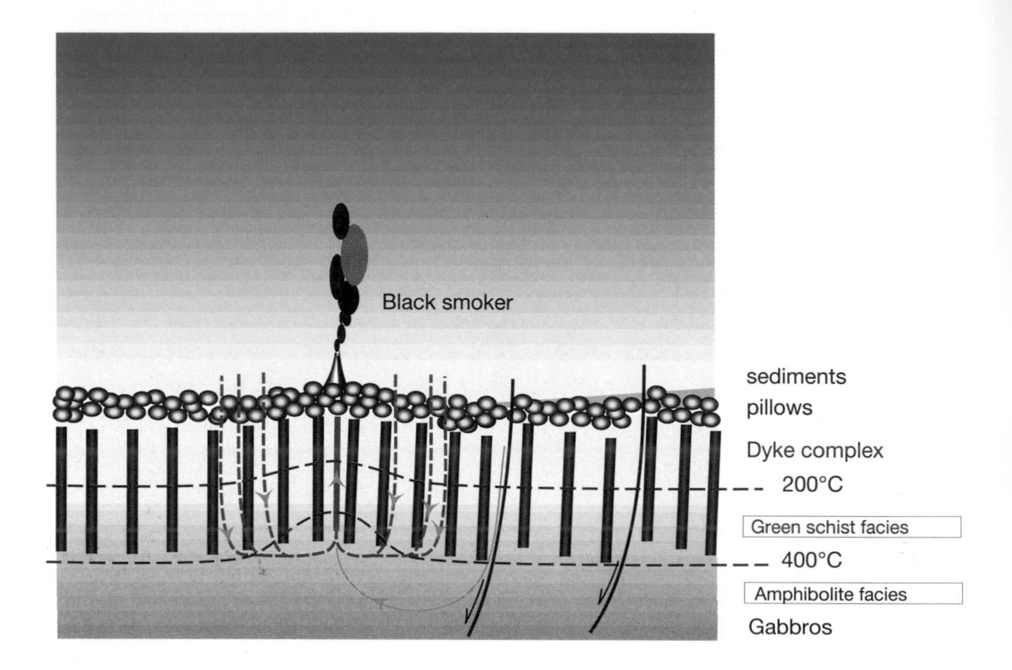

Figure 4.5
Hydrothermal circulation through a ridge or the "black smoker plumbing system". The main circulation (in *dashed lines*) turns around at the base of the dyke complex at a temperature of about 400 °C. The descent takes place along open faults and fissures parallel to the dyke complex and the ascent just below the ridge axis. A deeper circulation (in *light blue*) takes advantage of fewer faults to drain the gabbros

towards the ridge and starts to ascend, becoming progressively channeled towards the smokers. During this high-temperature circulation, the water alters and corrodes the rocks and dissolves the metals, which are then precipitated at the vent in contact with cold sea water. This alteration of rocks at between 450–350 °C is a type of metamorphism, more specifically the greenschist-facies metamorphism, because of the greenish tinge of the rocks and the faint schistosity acquired by them. As widespread penetration down beyond 2–3 km becomes more difficult, the waters here lead to a higher temperature metamorphism only in the direct vicinity of fractures. This represents the amphibolite-facies metamorphism which derives its name from the fact that in the gabbros it leads to the formation of amphibole, a hydrous silicate growing from pyroxene. The changes of this hydrothermal metamorphism in the oceanic crust with depth are shown in Fig. 4.5.

The hydrothermal activity of ridges is a fundamental phenomenon, the full implications of which are not yet fully understood. It is at the origin of certain mineral deposits, mostly of copper, which are precipitated at the vent of the hot springs and are deposited onto the ridge basalts. Such deposits are presently exploited in basalts of ophiolite sequences. The best known is located on the island of Cyprus and was already famous to the ancient Greeks, who named the island after the metal (cupros = copper). These deposits contain a total reserve of 1.5 Mt of copper. The manganese of the hydrothermal solutions travels farther and becomes fixed in the metalliferous manganese nodules covering vast areas of the ocean floors.

Mountains Covered by Carbonate Snow

Hydrothermal circulation through the entire ridge system, including circulation through the ridge flanks at lower temperatures, amounts to a total flux corresponding to 2 % of the annual discharge of all rivers of the planet. It suffices, however, to recycle the entire sea water through oceanic crust within a few million years. Such a flux of very hot and thus chemically aggressive water has major, but still poorly understood implications for the composition of the sea water and its control, and indirectly also for the composition of the atmosphere. We know now that hydrothermal activity extracts magnesium from sea water to fix it in the metamorphic rocks forming at depth. At the same time, it enriches the sea water in calcium dissolved from the same rocks. The concentration of these elements in sea water has practical importance for the crystallization and dissolution of calcium and magnesium carbonates, the main traps for CO2, thus contributing to the control of the CO2 content of the oceans.

In shallow water where the CO_2 content is rather low due to atmospheric exchange, carbonates crystallizing are fixed in the shells of marine organisms. Below a certain depth, the so-called lysocline, the CO_2 content increases and the shells start to become dissolved, to disappear completely below 5000 m, the so-called carbonate compensation depth (CCD). If we were able to scan submarine panoramas with our eyes, we would observe the submarine mountain ranges with their summits covered by a dirty snow of carbonates and the darker valleys lined by red deep sea clays. It is generally accepted that the depth of the lysocline depends on the supply of calcium and magnesium as well as on the CO_2 content of the sea water, i. e. it is controlled by the activity of the ridges and by the atmosphere due to exchange along the ocean-atmosphere interface. Thus the activity of the ridges might indirectly control the CO_2 content of the atmosphere, an equilibrium vital to mankind. This example illustrates the vast complexity and the unexpected ramifications of the "system Earth". We could say that man is at present running a large fan blowing CO_2 into the atmosphere, at the risk of fatally modifying its composition. Are the ridges controlling another fan by filtering the ocean water? Would they be able to restore the system to its equilibrium?

Did Life Originate on the Ridges?

Were the ridges the cradles of life on Earth? It appears possible that the physical conditions prevailing around the black smokers permitted the inorganic synthesis of amino acids and other prebiotic organic molecules. The most likely locus could have been small iron sulphide globules in the chimneys of the smokers. Around the smokers of the East Pacific Rise, very primitive bacteria were found living in a reducing environment up to a temperature of 110 °C, the highest temperature known to be tolerated by any organism. We must also understand that these oases are probably the only places on Earth where life does not draw its energy from the sun via photosynthesis in an oxygenated atmosphere (save for exceptional cases of oxidation in a reducing environment), but from the internal energy of the Earth made available by oxidation in a reducing sulphide environment. As the oxygenated atmosphere appeared rather late in Earth history, around 2300–2000 Ma ago, all earlier life forms had to be adapted to reducing environments and had to evolve in such an environment, which in this respect was comparable to the oases on the ridges. Considering that the hypothesis of an extraterrestrial origin of life and its introduction to the Earth by way of meteorites from outer space has been advocated, we may just as well propose an origin in cavities on submarine ridges.

5 Ophiolites – or in Search of Lost Oceans

Ophiolites, known to geologists for a long time, have acquired new esteem in view of the fact that they represent oceanic crust which we may visit on dry land. How is it possible that the first 10–15 km of this crust shear off and then creep along the floor of the ocean before climbing onto a continental margin? We shall find an answer when we investigate the best-preserved ophiolite belt, that of Oman on the eastern edge of the Arabian Peninsula. How are we to reconstruct the framework of an active ridge from the debris of this ophiolitic shipwreck? After a general discussion, we shall show that there are two main types of ophiolites and we shall see in the subsequent chapters that the two types are derived, the one from slow-, the other from fast-spreading ridges.

The Ocean, Dry-Footed

Why is there this interest in ophiolites, these formations appearing as green spots on geological maps? As outlined in the introduction, this is because they represent fragments of ocean floor which would otherwise be inaccessible. But how did these fragments run onto the borders of the continents? And finally, beyond a general definition, what do these ophiolites represent? We shall try to answer these questions in the following chapters.

a

b

c

d

Figure 5.1 a–d
The main ophiolite facies as observed in the Oman ophiolite belt. **a** Peridotites of the lower part representing the mantle. **b** Banded gabbros resulting from the crystallization in the magma chamber at the base of the crust. **c** Basaltic dyke of the dyke complex which feeds, in **d**, pillow lavas of the crustal cover

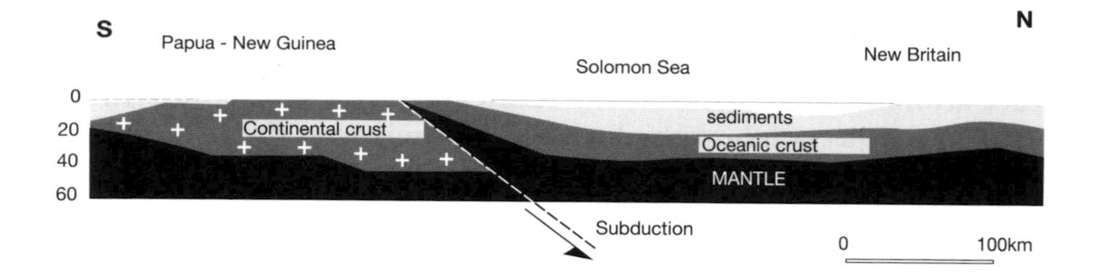

Figure 5.2
Thrusting of ophiolites in New Guinea. Geophysical model constructed from seismic and gravimetric data. (After D. M. Finlayson et al. 1976, Geophys. J. 44, 45–60)

Stranding or Obduction of Ophiolites on the Continents

We have seen that the usual destiny of the ocean floors is subduction into the mantle (Chap. 1). In contrast to this, geologists have proposed the term obduction, with the prefix "ob" for upward replacing "sub" for downward, describing the particular destiny of those ophiolites which during the advance of oceanic lithosphere against a continent end up stranded on the edge of this continent rather than disappearing in a subduction zone below it. There are a number of ophiolite occurrences which still bear witness to this unique behaviour. Some of them border the Pacific to the south and west like those of New Caledonia and Papua New Guinea. Figure 5.2 is a geophysical model illustrating the transport of the Papua ophiolites from the Solomon Sea. Seismic and gravimetric resolution here is insufficient to see that the eastern part of the island is covered by a few kilometres of rocks, the ophiolite forming a tectonic nappe pushed onto the island. It is nevertheless able to visualize the root zone of this nappe and to furnish an impression of the dimensions involved. We shall gain a better understanding of the importance of this geological phenomenon by retracing the history of the obduction of the Oman ophiolites. The model derived from the study of this ophiolite, however, is by no means unique, but as a general rule we may conclude that ophiolites are derived from very young lithosphere.

Ophiolites pose a great problem for which we shall propose a solution in the last chapter. In contrast to other manifestations of plate tectonic processes which we may observe in action in modern examples, we do not have available any example which clearly demonstrates how an ophiolite is "extracted" today from an ocean. This is rather deplorable, as whatever the quality of the geological observations used as a basis for the models of obduction, it is still difficult to imagine an event so gigantic in scale.

The Best Ophiolite in the World

Why should there be such interest in the Oman ophiolites to which we refer repeatedly in this book? To start off, it is the largest ophiolite complex in the world, extending for some 500 km along the Indian Ocean shore of the Arabian Shield (Fig. 5.3), covering it over a width of 50–100 km. The absence of vegetation in the desert climate assures spectacular outcrop conditions. And then, especially, in contrast to most other ophiolites, the convergence of the plates causing the obduction has not yet proceeded to its usual end, that is the collision of the continents carried by the plates. In geological terms we may predict that within about 2 Ma the Makran subduction zone at the edge of the European plate, with Iran representing the promontory will have swallowed the last 100 km of the ocean from which our ophiolite had been derived (Fig. 5.4f). There will then be a collision between the Arabian and the Eurasian plates, piling up a mountain range like the Alps and thereby erasing any traces of the ocean which originally separated these continental masses.The best ophiolite belt in the world will then become disrupted into slivers of different sizes, like the ophiolites of the western Alps.

Lying at present on the easten edge of the Arabian Shield, the Oman ophiolite was extracted tectonically some 100 Ma ago from the Indian Ocean. The ophiolites formed a nappe 10–15 km thick, a figure close to half the mean thickness of the continental crust. This nappe had travelled across its ocean of origin for several hundred kilometres before encountering the margin of the Arabian Shield. Under such a tremendous weight, the floor of the Indian Ocean was pushed down several kilometres. However, the top of the ophiolite may have come out on surface, forming a chain of islands. It is envisaged that during their marine travels in equatorial latitudes some of these islands were eroded to such an extent that they lost their 6-km-thick cover of oceanic crust, uncovering the peridotites of the mantle.

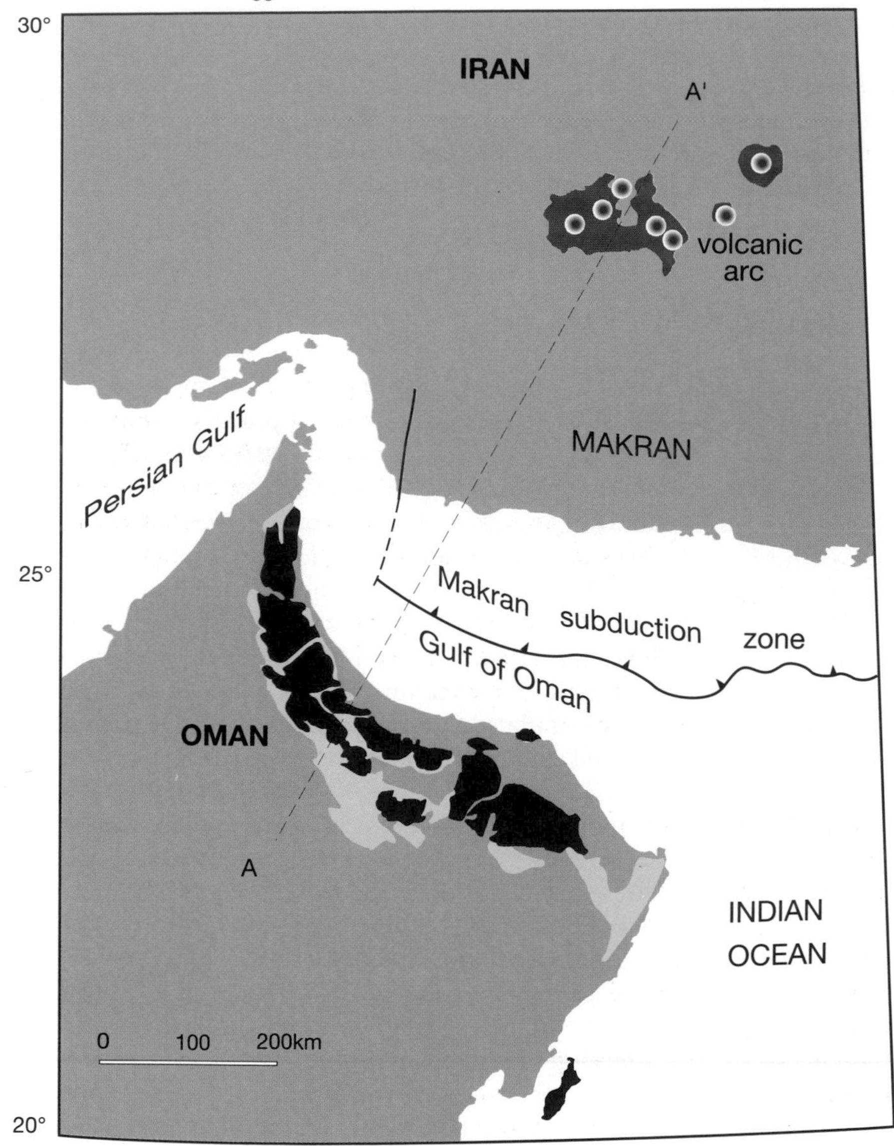

Figure 5.3
The Oman ophiolites (*deep blue*) and the Hawasina sediments (*yellow*), both thrust together onto the Arabian Shield. To the NNE, in the Gulf of Oman, the front of the Makran subduction zone is shown by the *thrust symbol*. Still farther north, in Iran, there is a volcanic arc genetically related to this subduction. Section *A-A'* corresponds to **e** in Fig. 5.4

The Long Journey of the Oman Ophiolites

The formations located at the base of the Oman ophiolites, like those of most other ophiolites, have recorded the history of the obduction and permit us to retrace it.

This history began with a reversal of the relative movements of the plates along the oceanic ridge which had given birth to the ophiolite. This was a rapidly expanding ridge, operating in the Indian Ocean up to the Cretaceous, 95–100 Ma ago. It was separated from the Arabian coast by the over 400-km-wide oceanic Hawasina basin. Because of factors to which we shall return in Chapter 9, expansion along the ridge was transformed into convergence over a geologically short period not longer than about one million year. The waning ridge served as the site triggering the transport (Figs. 5.4 and 5.5). It affected the rigid lithosphere which started to slide over the underlying hot and deformable asthenosphere. As the lithosphere is thinnest over the ridge, thrust will be initiated along the ridge or its vicinity once compression starts.

The moving lithosphere represents a wedge, the cutting edge of which is the axis of the original ridge and which becomes thicker farther away from it (Fig. 5.5). The transition between lithosphere and asthenosphere takes place in the mantle at temperatures of about 1000 °C. The slope of the interface separating lithosphere from asthenosphere is only a few degrees. The first convergent movements imprint themselves on the peridotites located along this surface. When these emerge from the mantle and come to rest upon the oceanic crust over the western flank of the ridge, they start to heat these crustal rocks like a flat iron to nearly 900 °C. These rocks start to recrystallize while being deformed under the moving load of the overlying lithosphere, thereby being subjected to intense metamorphism. The peridotites lose their heat to their substrate and cool rapidly, and at around 850 °C they become rigid while still very close to the axis. The sliding of the ophiolite nappe imprints itself now in the underlying oceanic crust which remains deformable at these temperatures. Heat and deformation transform the basalts of the crust into high-grade metamorphic rocks, viz amphibolites.

During its oceanic travels at a rate of several centimetres per year, the ophiolite nappe loses heat. The temperature at its base will now have dropped to about 400 °C. It slides over a veneer of basalts and radiolarites of the underlying crust which become transformed into low-temperature metamorphic rocks, viz the greenschists. The traces of the movement towards Arabia are now imprinted onto these greenschists. Like the blade of a bulldozer, the front of the nappe

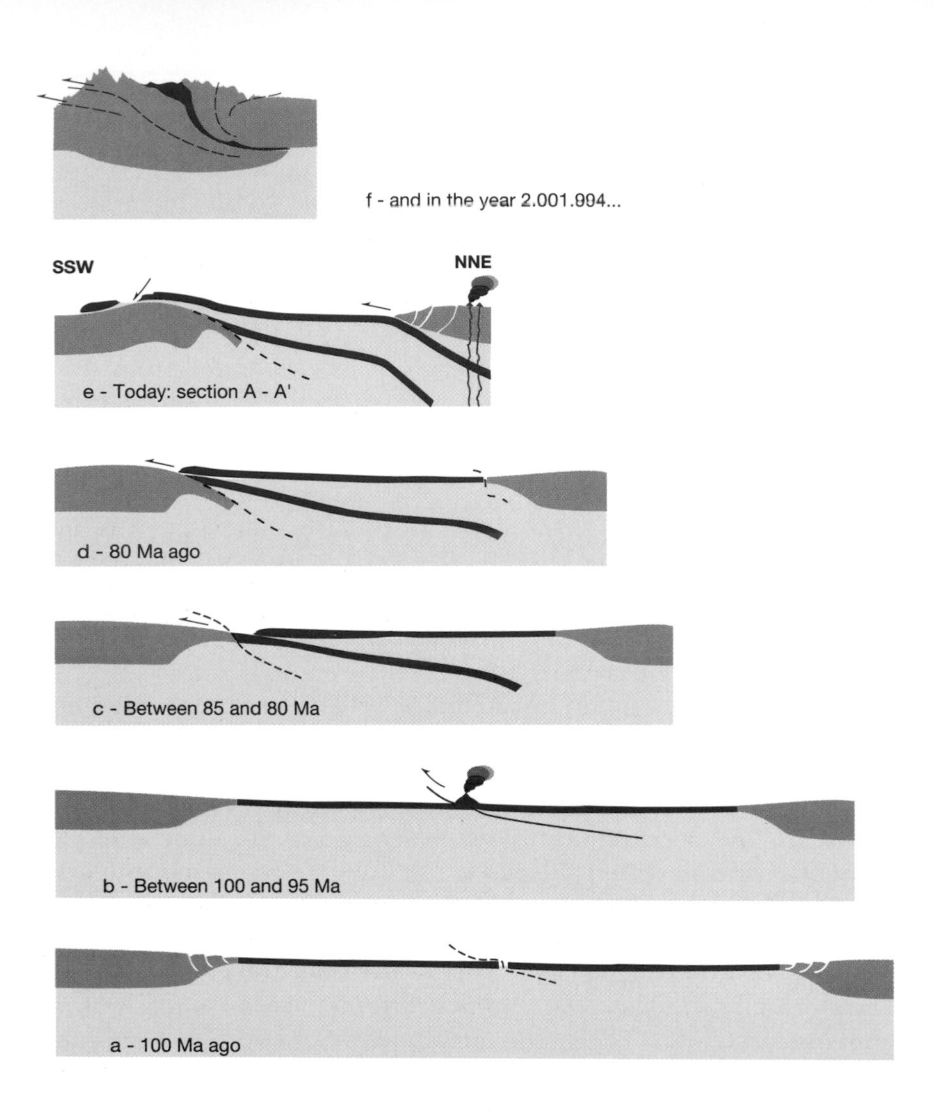

Figure 5.4 a–f
Profiles illustrating the obduction of the Oman ophiolites from their formation about 100 Ma ago (**a**) to the present stage (**e**), along line A-A' in Fig. 5.3. (**f**) The collision about 2 Ma from now in the future is shown. (After A. Nicolas 1989, Structures of ophiolites and dynamics of oceanic lithosphere, Kluwer Ed., 367 p.)

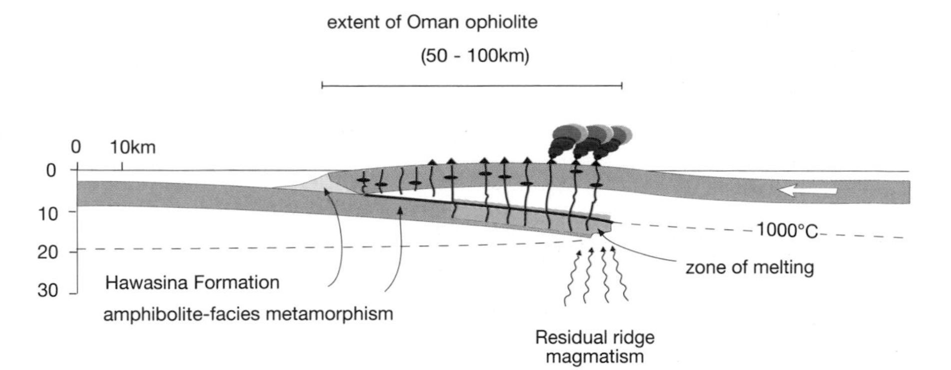

Figure 5.5
Thrusting of the ophiolites onto the ridge where the lithosphere is thinner and weaker. The "ironing" effect of the warm thin lithosphere leads to amphibolite facies metamorphism and to incipient melting of the crust (*shaded portion*). Mixing of the melting products with basalts resulting from the residual activity of the ridge could be responsible for the island-arc-type volcanism. (After F. Boudier et al. 1988, Tectonophysics, 151, 275–296)

pushes along all formations deposited on the oceanic floor of the Hawasina basin as well as any structures built upon it, sediments of increased age and thickness as the continent is approached, guyots and volcanic islands fringed by coral reefs and eventually the rocks of the continental margin encountered close to shore. These mixed formations are presently observed in front of and under the Oman ophiolites (Fig. 5.6).

Around 85 Ma ago, after some 10 Ma of marine travel, the nappe reached the Arabian Shield and its weight caused the edge of the continental margin to bend downwards and to slide under the ophiolites (Fig. 5.4). This incipient subduction of the margin became progressively slowed down because of the resistance of the lighter continental crust to burial under heavier oceanic lithosphere. This resistance made itself felt also by beginning horizontal movements within the continental margin itself. These movements led to thickening of the margin and made it even more resistant to subduction which petered out 100–200 km from the margin to a depth of some 50 km. We have already shown in Chapter 1 that the density differences encountered may explain the resistance of continental plates to subduction and, as a consequence, their persistence on the surface of the Earth.

Figure 5.6
The front of the Oman nappe. In the *top photo* we recognize, *from right to left,* peridotites exhibiting a near-horizontal low-temperature foliation, then in *darker shade* the metamorphic formations of the base, and at the *far left,* the underlying sediments of the Hawasina Formation. These rest upon the domed limestones of the margin of the Arabian Shield which occupy the entire rear-ground of the photo. The *bottom photo* shows an exotic block (in *white*) below the edge of the nappe in the rear-ground. In the latter, the Moho is clearly visible, separating the light-coloured peridotites from the gabbros forming the dark layers capping the summits

The subduction of the Arabian Shield came to a halt under the ophiolites after an only modest advance considering the dimensions involved. This led to two consequences:

- The convergence between the Arabian and the Eurasian plate could not be held up, as it was part of a global tectonic process leading to the opening of the Atlantic and the drift of Africa towards Eurasia (Fig. 1.4), a new process of subduction was started, this time along the southern edge of the Eurasian Plate in the Makran subduction zone (Figs. 5.3 and 5.4). By indirect evidence, e. g. dating the intrusion of granites resulting from island arc activity along this subduction zone, we know that this zone became active very soon after the obduction of the ophiolites onto the Arabian Shield. We have seen that the zone remains active to the present time, heralding a sombre destiny for our ophiolites.
- Secondly, the blockage of the Arabian margin below the ophiolitic nappe. Freed of the traction at the base by the subduction now taking place in the Makran zone, the margin started to rise and to form a dome which favoured the eventual sliding of certain parts of the nappe towards the west into their present position. East of this dome, the oceanic lithosphere which had piled onto the continental margin also started to slide but eastward. The rock formations of this margin, which are presently observed along the Indian Ocean coast, were thus denudated. These high pressure metamorphic rocks display features related to their burial of up to 50 km during the subduction process.

The Ophiolites as Fragments of Young Oceanic Lithosphere

We have at our disposal relatively precise geological descriptions of some 100 ophiolite massifs around the world. Whatever the recent state of tectonic arrangement or disarray , most of them exhibit at their base a metamorphic layer which is the result of their emplacement as a hot layer on the ocean floor, as in the Oman ophiolite massif. In order to exert the "ironing" effect onto the ocean floors over which they passed during their travels, the ophiolites had to be hot themselves. We know furthermore that the thickness of most ophiolites does not exceed 10–15 km, which would correspond to the thickness of the original lithosphere and we have seen that lithosphere rapidly cools and increases in thickness with age (Chap. 1). Thin and hot lithosphere must thus be very young, the best example in this respect still being represented by the Oman ophiolites where the ophiolite-to-be had been derived directly from the axis of an oceanic ridge, or its immediate vicinity.

Based as it is upon theoretical considerations, this conclusion is corroborated by geochronological data available on a number of ophiolites. There is thus no significant age difference between the last rocks formed on the axis of the Oman ridge and the first signs of metamorphic recrystallization induced by the hot peridotites travelling over the underlying oceanic crust. This shows that in this particular case the thrusting has directly interrupted the activity of the ridge. The reversal from expansion to compression was produced quickly and is hardly discernible by the available geochronological methods, the precision of which is in the range of 1–2 Ma.

This is not always the case and in certain ophiolite belts age data show that transport started several million or tens of million years after the formation of the ophiolite rocks. Some scenarios try to explain this transport as resulting outside a ridge context in lithosphere remaining fairly young. Because of the difference in relief between two flanks, transform faults may favour the transport of the relatively higher flank (Fig. 3.6). The energy actually required to transport this lithospheric portion here would be reduced because of its already elevated position.

Comparative "Ophiolitology"

Ophiolites Do Not Belong to a Unique Species

Reinstated by plate tectonics, which itself resulted from marine discoveries, the study of ophiolites has advanced considerably over the last 20 years. Whereas the "genus" ophiolite is well defined by now, we know also that it includes a number of "species". Ophiolites may be distinguished from each other by the exact nature and the relative position of their constituent units. Thus in the Xigaze ophiolite of Tibet, the gabbros are virtually non-existent, whereas in others, like those of Oman, they are present in a thickness constantly above 2 km. The ophiolites may furthermore be distinguished by their internal structures. The Xigaze ophiolite, for instance, is characterized by a dyke complex which is not vertical but horizontal; and last, they are distinguishable by their chemical composition or, dealing with the Earth, geochemistry. Thus the Canyon Mountain ophiolites of Oregon possess a crust of andesitic instead of basaltic composition. This corresponds to a relative enrichment of silica, alumina, calcium, alkali elements and water against most other ophiolites.

In biology, comparative anatomy permits us to establish a correlation between the structure of certain organs and the environment in which the organisms of a single group evolved. Cuvier brought this method to perfection in his study of the evolution of horses, as a function of their environment, based purely on fossil bones.

In the same way, the comparative study of ophiolites, "comparative ophiolitology", aims at making use of the wealth and variety of ophiolites known for defining the original oceanic environments. Under the term oceanic environment, we may understand geodynamic as well as physical ones. To give an example of the former situation: the centre of expansion, the cradle of the ophiolite, may belong to an oceanic ridge or a basin above a subduction zone, two vastly different geodynamic environments. It is important to distinguish between these if one is to reconstruct the original oceanic environment in paleogeographic syntheses. In the second situation, one would look for a definition of the physical structure of the ridge. Was it a slow- or a fast-spreading ridge? Before we embark on this distinction, let us take a closer look at the first case.

The Geodynamic Environment of Ophiolites

As the island arcs and the back-arc basins located above subduction zones represent relatively unstable sites and are frequently situated close to the continents, it has been proposed that most ophiolites have been derived from such localities. It is actually easier to imagine the emplacement of an ophiolite onto a continental margin if its source was in an oceanic region close by and if it is kept in motion by processes which may well degenerate into thrusts than if it was formed on an oceanic ridge far removed, the operation of which is moreover rather continuous. These general ideas were reenforced when it was realized that a geochemical signature referred to as typical of island arcs could be noted in a number of ophiolites. What does this tell us?

Submarine lavas which have been dredged or cored in different environments exhibit chemical compositions or geochemical signatures which differ slightly. On the basis of the oxydes of the eight main constituting elements, i. e. silica, alumina, iron, magnesium, calcium, sodium, potassium and titanium, the basalts are rather similar, except that in the island arcs we also encounter andesites. Whereas the concentrations of these major elements in the lavas are of little discriminating value, the contents of the trace elements chromium, zirconium, yttrium and the rare earths are more promising. The most pronounced differences are developed between the lavas of oceanic ridges (MORB = Mid-Ocean Ridge Basalt) and those from island arcs and back-arc basins. In contrast to the ridge lavas, those of the arcs and back-arc basins are derived from the melting of mantle foundering below a subduction zone. This melting is induced by the ascent of fluids, and mainly of water derived from the subducted lithosphere, which at a depth of about 100 km starts to dehydrate and recrystallize intensely. The lavas resulting from these processes will reflect this particular context.

Because this typical signature or fingerprint has been observed in the lavas of a number of ophiolites, their formation in an arc environment has been proposed. This interpretation is irrefutable for an ophiolite like the one at Canyon Mountain already referred to. Its original crust exhibits an andesitic composition similar to that of island arcs. However, it is by now clear that this interpretation has been extended too liberally to other occurrences. It has been shown, e.g. in the Oman ophiolite, that the earliest stage of transport was accompanied by the extrusion of lavas strongly contaminated during their upward ascent through the over-run crust (Fig. 5.5). As these lavas have thus acquired an arc affinity in this particular way, the attribution of the entire ophiolite to such an environment is no longer tenable. A better insight into the geochemical factory of the ridge where the Oman ophiolite originated reveals that, aside from the dominant MORB, back-arc or arc signatures may eventually be generated.

The Functioning of the Ridges: a Structural Perspective

As we want to understand the functioning of the oceanic spreading centres, we have to interest ourselves mainly in the message provided by the ophiolites about their physical environment of formation. The Fiji ridge at present opens up a back-arc basin to the west of the volcanic arc of the Fiji islands and above the Pacific plate, which here is subducted to the west. Recent dives by the submersible NAUTILE confirmed that this rapidly expanding ridge is physically similar to the ridges of the open oceans which expand at similar rates. The questions of the geodynamic environment thus does not appear to affect our research. Because of this, in the following text, the terms ridge or rift will refer to oceanic spreading centres without particular reference to their geodynamic situation, unless specifically mentioned.

The ophiolites may assist us in a physical and structural perspective to understand the operation of ridges provided we ask the proper questions. What type of structural or dynamic framework may we assign to oceanic ridges from our ophiolite studies? How could we dress the model of a ridge, which is outlined in rough strokes by geophysics with the "clothing" frequently scattered throughout the ophiolite massif? How could we fill the model with life?

The Ophiolite Puzzle and Its Key

Stranded on the edge of the continents like the hull of a ship, the ophiolites are usually dismembered if not dispersed by the wave carrying them. Their condition becomes even worse when they are caught up in the vice of continental collision frequently following their obduction (Fig. 5.4). It is the task of the geologist to "refloat" the ophiolite or at least to bring the various fragments mentally into order again. This operation is indispensible if we are to reconstruct the oceanic spreading centre at which the ophiolite was formed, be this a rift or a slow- or a fast spreading ridge.

However, we are not deprived of all means to carry out this job. We know, for example, the relative position of the different units: the mantle peridotites below the gabbros of the lower crust, etc. We also know that the Moho, which separates the peridotites from the gabbros, was originally horizontal in the lithosphere. Measuring the orientation of this surface in the field, we will be able to reorient to the horizontal position an ophiolite that had been tilted (Fig. 5.7); and we know moreover that the basaltic dykes of the dyke complex which are remarkably parallel to each other (Fig. 5.1c) intrude because of the separation of the two sides of the original ridge or rift. If we repeat this intrusion on a metre scale, the mean width of each dyke, over several tens of kilometres, we obtain the principle of emplacing bands of magnetic anomalies with opposing field orientations parallel to the ridge. The dykes making up this complex are thus vertical and should consequently be arranged at right angles to the Moho, thereby facilitating a control for any rotation that might have affected these rocks. However, it is more interesting to note that their strike is parallel to that of the original ridge (Fig. 5.7). In an ophiolite the mean strike of the dykes in the dyke complex thus indicates the orientation of the original ridge or rift.

If the various elements of an ophiolite are not scattered too widely, we shall be able to reconstruct the structural framework of its original oceanic spreading centre (horizontal plane, direction of the ridge).

Figure 5.7 a–g
From the field to the model ▶

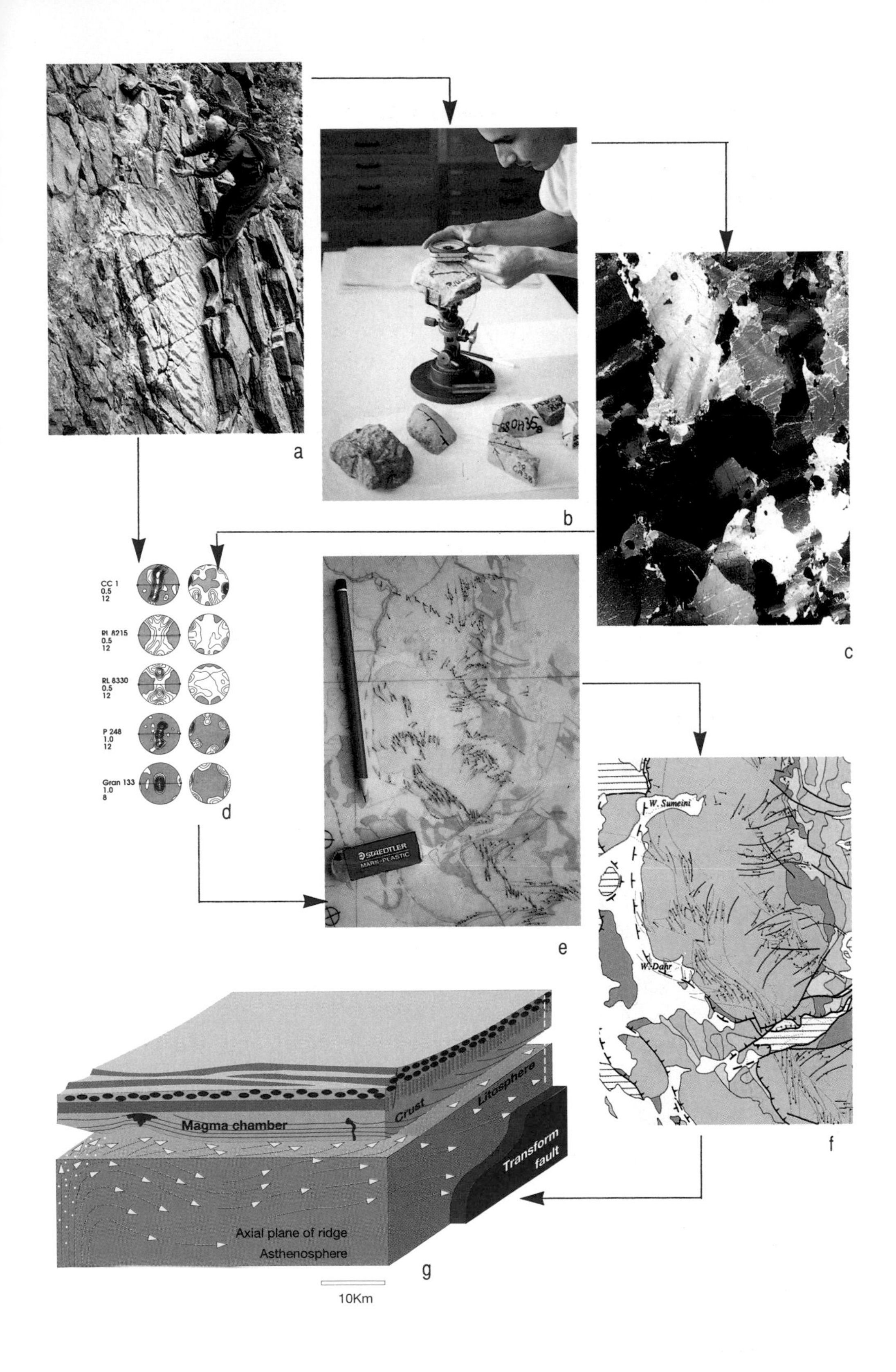

a
b
c
CC 1
0.5
12
RL 8215
0.5
12
RL 8330
0.5
12
P 248
1.0
12
Gran 133
1.0
8
d
STAEDTLER
MARS-PLASTIC
e
W. Sumeini
W. Dahr
f
Crust
Litosphere
Magma chamber
Transform fault
Axial plane of ridge
Asthenosphere
g
10Km

The Two Types of Ophiolites

It might appear strange that, after having stressed the wealth and diversity of ophiolites, we should attempt to reduce them to just two types. In doing so, we neglect the geochemical differences which could be symptomatic of the original geodynamic environment and we are not looking for a chance to profit from peculiarities which tie the origin of the ophiolite in question to a particular locality on the ridge, be this a fracture zone or an overlapping spreading centre.

In the following chapters, the comparison between ophiolites and ridges will show that our two ophiolite types usually correspond either to fast-spreading ridges like those in the eastern Pacific or the Fiji back-arc basin or to slow ridges like that in the Atlantic. As a model of the former type we shall take recourse again to the Oman ophiolites, the best representative of the type encompassing most other ophiolites. It includes some of the best-studied like those of New Guinea, New Caledonia, the Philippines, Cuba, Newfoundland, Cyprus and numerous occurrences in Turkey. The Trinity ophiolite of California is a model of the second type which, however, is less widely distributed on the surface of the Earth. There are obviously also ophiolites which occupy an intermediate position between these two types. One of these is represented by the Xigaze ophiolite in Tibet which was mentioned at the start of this chapter.

This analysis, based solely on spreading velocity, may, under certain circumstances, hide another source of differences related to the existence of distinct source mantle among oceanic ridges and, thus, ophiolites. It is admitted here that thermally the mantle is ideally stratified, meaning that at a given depth, the temperature is everywhere the same. This is in contradiction with the presence, below the hotspots described in Chapter 2, of plumes which are hotter than the surrounding ideal mantle. Consequently, we should add to our distinction between slow- and fast-spreading ridges, another between colder- and hotter-spreading ridges. Thus, in the northern Atlantic, the Reikjanes Ridge, south of Iceland and under its hotspot influence, displays features distinct from more southerly and colder ridge segments. We believe, however, that in our general analysis this effect can be ignored.

Lherzolites and Harzburgites

We are referring to the two types of ophiolites as harzburgitic or lherzolitic, depending on the nature of the peridotites making up their mantle component. An excursion into the mineralogy of peridotites is useful here to understand better the role played by the mantle in oceanic expansion.

The normal upper-mantle peridotites are called lherzolites. This is a rock which, like all peridotites, is dominated by olivine, but also contains up to 30–40 % of two other minerals belonging to the pyroxene group. One of these pyroxenes, enstatite, is chemically similar to olivine but has a slightly higher silica content, whereas diopside, the other pyroxene, is chemically much more varied. In addition to silica, and to the iron and magnesium of the above-mentioned minerals, it also contains calcium, aluminum, sodium and chromium. In a gross oversimplification, we could say that the composition of a diopside approaches that of a basalt. In addition to the minerals mentioned, the lherzolite contains smaller amounts (below 10 %) of alumina-rich minerals, the mineralogical nature of which varies with the depth at which the lherzolite equilibrated with the mantle. Below 75 km the mineral will be garnet, between 75 and 30 km spinel, and above 30 km plagioclase. The presence of one of these three minerals in a lherzolite allows us to estimate its depth of formation or, more precisely, the depth at which the peridotite in the ascending mantle had cooled sufficiently to freeze its composition (Chap. 7).

The lherzolite is the source rock in the mantle from which basalt is extracted by partial melting. Melting will primarily attack the aluminous accessory minerals and the diopside which, with increasing degree of melting, become progressively rarer in the lherzolites. The refractory residue of this melting is the harzburgite which in the process becomes relatively enriched in olivine and enstatite, as these minerals are less susceptible to partial melting. The harzburgite still contains an accessory amount of spinel which, however, is depleted in alumina due to the preceding melting and correspondingly enriched in chromium. Basalt coming from depth and traversing the harzburgites located just below the Moho may dissolve their enstatite. The resulting rock, consisting only of olivine and the chromium-rich spinel chromite, is referred to as dunite. The mineralogical composition of the main types of mantle peridotites is compiled in the following table.

	Mineral			
	Olivine SiO_4 $(Mg_{0.9}\,Fe_{0.1})_2$	Enstatite Si_2O_6 $(Mg_{0.9}\,Fe_{0.1})_2$	Diopside Si_2O_6Ca $(Mg_{0.9}\,Fe_{0.1})$	Minerals containing Aluminum
Peridotites				
Lherzolite	60–70%	20%	5–10%	5–10% Plagioclase <30 km 30 km$<$ Spinel<75 km Garnet >75 km
Harzburgite	70–80%	20%	0–5%	Spinel 5%
Dunite	95%			Spinel (Chromite) 5%

To sum up the table presented above, let us keep in mind that lherzolite is the normal upper-mantle peridotite. Its mineralogy shows that it can produce basalt by partial melting. The residue of this melting is a peridotite called harzburgite. We may thus write the following reaction:

lherzolite → harzburgite + basalt.

The harzburgite thus is a refractory rock which will no longer give rise to basalts. However, alongside the olivine it contains enstatite, a silicate which may be dissolved by magma rising from greater depths. The resulting solid residue, the dunite, is the most refractory rock known. We may write another reaction:

harzburgite + basalt (1) → dunite + basalt (2).

Comparison of the Characteristics of the Two Types of Ophiolites

In Fig. 5.8, we compare, in vertical columns, structure and composition of the Oman and the Trinity ophiolites which we have selected to illustrate the two different types. Let us now describe in detail the features by which these two types differ from each other.

Peridotites: These are harzburgites in the Oman ophiolites and lherzolites at Trinity. In the upper portion of both massifs, we observe irregular bands of dunite which, however, are somewhat more abundant in the Oman ophiolites. The Trinity lherzolite contains plagioclase, which indicates that this rock "froze" its present mineralogy in the mantle at a depth of less than 30 km. We note moreover distinct orientations of the high-temperature deformation planes which resulted from

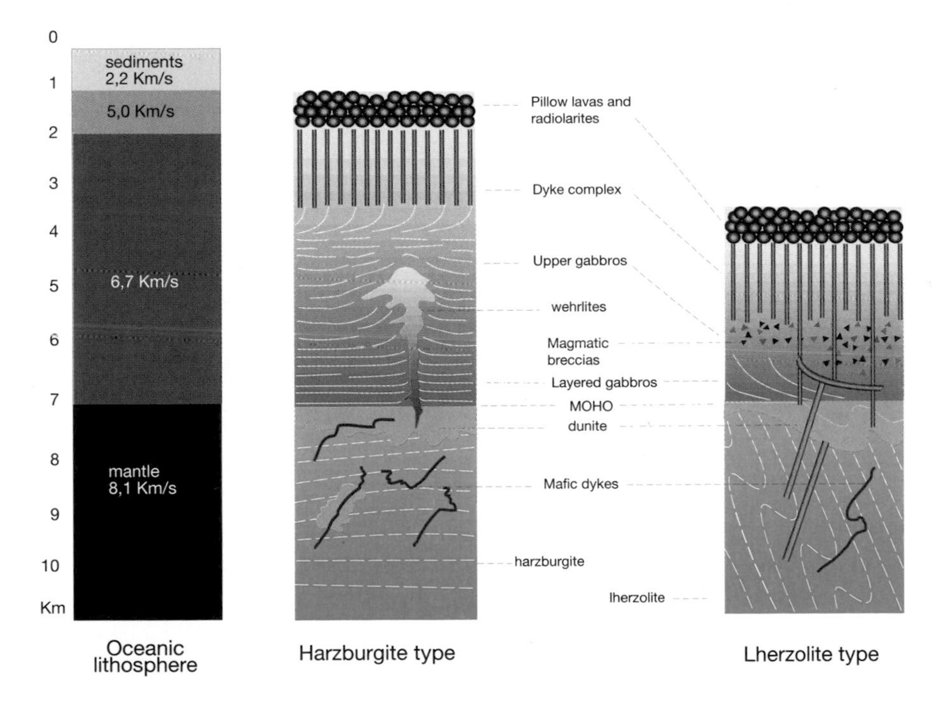

Figure 5.8
Columns comparing the structure of the oceanic crust as defined seismically with the two main types of ophiolites: the harzburgite type as illustrated by the Oman ophiolites and the lherzolite type, by the Trinity ophiolites of California. (After F. Boudier and A. Nicolas 1985, Earth Planet. Sci. Lett., 76, 84–92)

movements in the asthenosphere below the ridge. In Oman these planes are horizontal or only slightly inclined, whereas at Trinity the planes are much inclined except for the last few hundred metres below the Moho, where they tend to turn horizontal. We shall come back to this observation in Chapter 8.

Gabbros: The gabbros exhibit a more pronounced layering (Fig. 5.1b), and are thicker and more continuous in the Oman ophiolites than at Trinity. In the latter, the transition between the peridotites and the dyke complex takes place over thin gabbro lenses which are difficult to identify as they have been deformed plastically as well as recrystallized and fractured under the influence of hydrothermal fluids. It is mainly the reduced thickness of the gabbro unit which accounts for the lower crustal thickness at Trinity (2–3 km) as compared to 6 km in Oman and in the oceanic crust in general.

Dyke complex, lavas and intrusions: In these units, there appears no basic difference between the two types of ophiolites. However, at Trinity, the dykes are not restricted to the level of the dyke complex and they extend down into the peridotites. In contrast to this, the crust of the Oman ophiolites is locally invaded by intrusions of wehrlite, a black rock rich in olivine crystals set in a gabbroic matrix. These intrusions are highly diverse in size and shape and intruded the gabbros when these were still hot. This is proven by the fact that the wehrlites deform the gabbros in a still plastic state and by the large crystals developed in the contact zone between the two rock types. An igneous intrusion in a cold environment would be characterized by straight fractures along which magma would intrude and then freeze. This is typically observed in the dyke complex.

Let us return to the two columns of Fig. 5.8. The Oman ophiolites as a representative of the harzburgite type possess a mantle component which is notably more depleted in basalt than that of the Trinity ophiolites, the typical example of lherzolite type. The composition of the latter is thus closer to that of the original mantle. Broken lines in the mantle portions of the two ophiolites represent the fossil traces of planes of movement in the mantle below the original ridge at very high temperatures. We observe that these planes are practically horizontal in Oman and vertical at Trinity. Above this, the crustal portions of the two occurrences differ mainly at the gabbro level. These are well layered and thick in Oman and reduced in thickness, discontinuous, deformed and in disarray at Trinity. This results in the fact that in the latter locality the crust only attained half the normal thickness of the oceanic crust.

6 Mantle Metallurgy

Under their bleak cover, the peridotites of the mantle retain the secrets of the functioning of ridges. During their movements below the ridges, they deformed like a white-hot metal ingot between two rollers. The analysis of these "metallurgical" deformations permits us to reconstruct the motions of the mantle under the ridges. During its ascent, the mantle becomes subjected to partial melting under the ridges, and around a depth of 50 km the basaltic liquid thereby produced creates a pressure that suffices to fracture the peridotites and starts to intrude towards the surface. Within a few weeks only, these powerful intrusions push the oceanic crust apart by 1 m. These intrusions are reactivated at intervals controlled by the rate of expansion of the ridges which covers several such intrusions per century. They are an expression of the basic episodic nature of volcanism.

The Message of the Mantle

Ophiolites Deprived of Their Mantle

Bald Mountain, as put into music by Mussorgsky, or Red Mountain on maps of English-language countries, Kizil Dag (Red Mountain) in Turkey or Jebl Aswad (Black Mountain) in Arab-speaking countries – these far from encouraging terms introduce peridotite massifs to the traveller. The denudation of these blackish or reddish mountains is related to their particularly depleted chemistry dominated by iron, magnesium and silica. Black and even austere, these massifs are hardly more attractive to the geologist than to the layman. However, despite the monotony of their landscapes and the brown uniformity of their outcrops, they harbour the secrets of the terrestrial mantle.

Figure 6.1
Peridotites in the field. The only structure well developed in these otherwise featureless rocks is a banding usually marked by a relative enrichment or depletion of olivine with respect to the other minerals (mostly pyroxenes). The banding results from intense deformation which flattens and draws out into parallel lenses any heterogeneities originally present in the peridotite. As a result, an olivine cluster or a gabbro dyke are transformed by the deformation into centimetre-thick lenses extending over several metres

Geological studies of ophiolites were long restricted to their crust components. The gabbros, dykes and lavas have been gone through with a fine-toothed comb, whereas the vast expanses of peridotites, although much better developed in the field, were shown on geological maps mainly by a single colour only. Mineralogical, petrological and geochemical studies of spot samples taken across the peridotite massifs contributed much to our knowledge of the chemistry, physical conditions and mantle dynamics in general, as well as of the origin of ophiolites themselves. However, in the absence of a structural framework based on systematic field studies, the results of spot samples do not necessarily form a recognizable pattern. In analogy to opening a safe, it is not enough to know the individual numbers, but also their correct sequence. In studies of the mantle we were thus lacking the structural key, which was difficult to find, as at first sight the peridotites are desperately monotonous, save for discrete banding which we may follow from spur to spur in the hills (Fig. 6.1).

Mantle Metallurgy

The key was eventually discovered some 20 years ago by geologists and physicists cooperating under the auspices of the French research institute CNRS. We had observed that under the optical microscope, olivine, the dominant mineral in peridotites, exhibits deformational structures which are astonishingly similar to those of metals forged at high temperatures or pressed in a rolling mill. Such materials, which may undergo intense deformations at high temperatures in an entirely solid state, are called plastic. Optical and electron microscope studies of olivine deformed in the mantle or in the laboratory at temperatures above 1000 °C confirm this analogy with metals. This led to the "metallurgical" analysis of peridotites and to introducing to the study of the mantle the information which physicists had acquired in the field of metallurgy. This enlightenment has permitted us to trace the movements of peridotites in the mantle. It was extended rapidly also to the crustal rocks, leading to the analysis of movements referred to as kinematics analysis. This has deeply influenced structural geology which, only descriptive and static until then, entered the new field of geodynamics with kinematic analysis.

An olivine crystal is deformed like a deck of cards in which the cards slide over each other (Fig. 6.2). A peridotite is made up of an assemblage of olivine and other minerals which under the influence

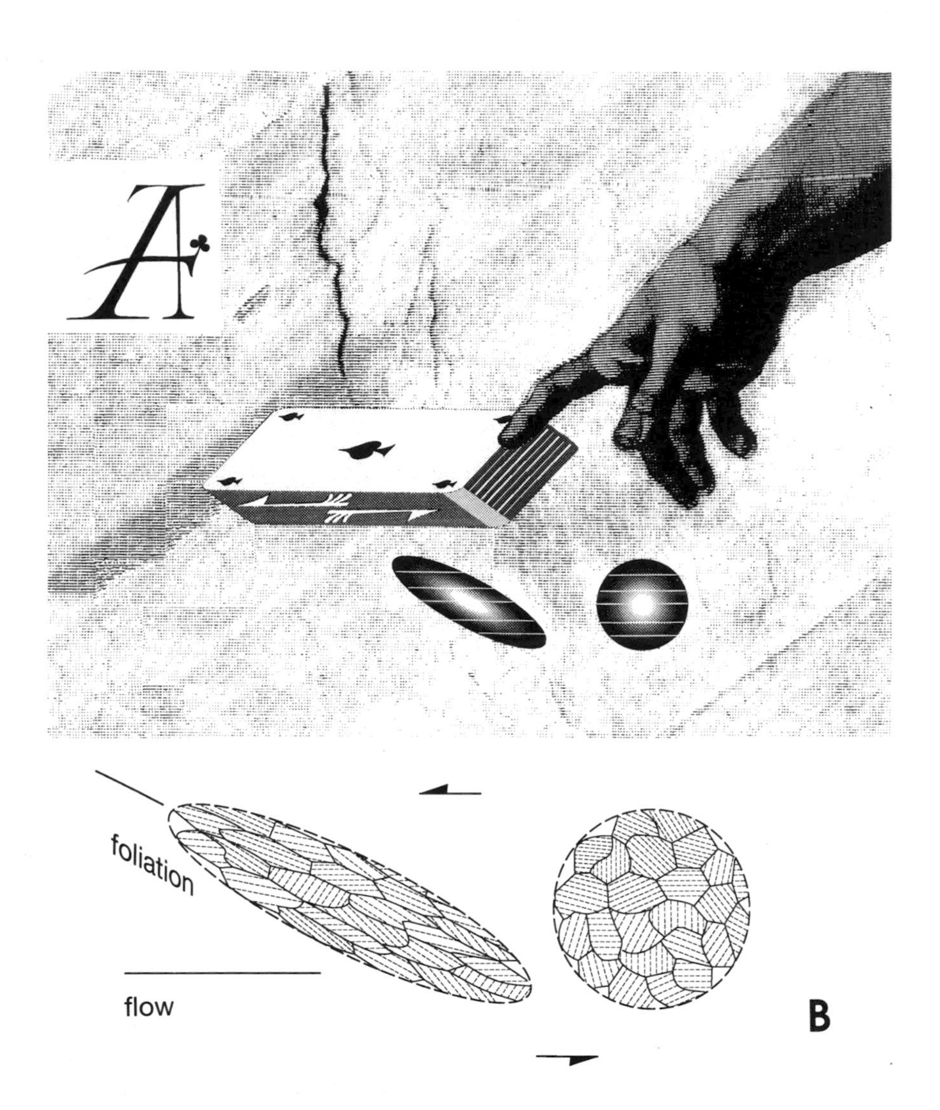

Figure 6.2 A,B
Minerals are deformed like the deck of cards by the finger of God (Michelangelo, Sistine Chapel), gliding along planes and lines controlled by the properties of the crystals. A *circle* drawn on the side of a mineral is transformed into an ellipse by gliding (**A**). This also applies to the rock itself (**B**). The deformation of the rock takes place through sliding controlled by its constituting minerals. The mean direction of flattening of the minerals determines the foliation, here obligue to the direction of shearing (*arrows*). When the degree of deformation is high, as commonly in peridotites, the two directions approach each other and may not need to be distinguished. (After A. Nicolas 1984, Principes de tectonique, Masson Ed., 223 p.)

Figure 6.3
Foliation or the plane of mineral flattening in a peridotite

of an oriented pressure tend to flow plastically and to assume a common orientation and to respond as a whole like the individual olivine crystal or the deck of cards (Fig. 6.2). This sliding layer by layer induces into the rock a planar orientation of the minerals, the so-called foliation (Fig. 6.3), whereas the direction of the flow is recorded by the elongation or stretching of minerals, the lineation. These foliations and lineations are recorded in the field, and by averaging a number of these measurements in the laboratory, we are then able to reconstruct the planes and directions of plastic flow in the rocks. We shall state here that the plane of foliation coincides with the plane of flow and the lineation with the direction of flow, although close scrutiny of Fig. 6.2 shows that this is not always strictly the case.

In the light of our knowledge of metallurgy, the study of peridotite structures under the microscope permits us to define the physical conditions prevailing during plastic flow. It is fairly easy to distinguish in ophiolitic peridotites structures resulting from oceanic transport at temperatures of 1000–900 °C (cf. Chap. 5) from structures formed at notably higher temperatures (1250 °C and above) and produced during the ascent of the material under the original ridge (Fig. 6.4).

Figure 6.4 a,b
Thin sections of peridotites under the polarizing microscope. Foliation is oriented E-W and the longer side of the slide measures 25 mm. Disregarding the bright interference colours which here are without significance, we note the difference between a peridotite deformed at 1250 °C in the presence of traces of basaltic liquid in the asthenosphere below a ridge (**a**), and the same type of peridotite deformed at 1000 °C under lithospheric conditions during thrusting onto the ocean floor (**b**)

The former type of deformation is referred to as lithospheric, as flow here affects the lithosphere, whereas the latter is called asthenospheric, as flow takes place in the asthenosphere rising under the ridge.

Moving Mantle Under the Ridges

Based on our theoretical knowledge gleaned from the study of ophiolitic peridotites, we shall now set the mantle below an oceanic ridge into motion.

This attempt is well within our reach, as in a little-deformed ophiolite massif we can reorient the floor of the ocean and the base of the crust – the Moho – back to horizontal and we know the direction of the ridge from the orientation of the dyke complex (Chap. 5). We have learned to recognize the imprints of deformation, viz. lineation and foliation, and we can distinguish deformations of peridotites by flow at very high temperatures in the asthenospheric mantle under the ridge from flow of the already cooled or still cooling mantle. After locating in a geological map the brown colour indicating the peridotites of an ophiolite belt, we can go to the field and systematically measure the foliations and lineations imprinted on these rocks by the asthenospheric deformations. Once the fieldwork supported by laboratory studies is achieved, we shall be able to draw a map of the mantle movements in the vicinity of the ridge from which the ophiolite originated (Fig. 7.1).

A Rule of Conduct

From maps of mantle movements which are available by now on some 15 ophiolite belts around the Earth, we shall reconstruct in Chapter 8 the strange round-about of the mantle under a ridge. Before this, however, we have to decipher the message hidden in the ophiolites.

An ophiolite is a fragment of oceanic lithosphere which formed under a ridge, and drifted over a certain distance prior to becoming uprooted by obduction from its origial natural site (Fig. 6.5a). At some distance from the ridge, the movement of the asthenosphere consists of sliding along planes that are parallel to the overlying rigid lithosphere. Then the upper part of the asthenosphere cools pro-

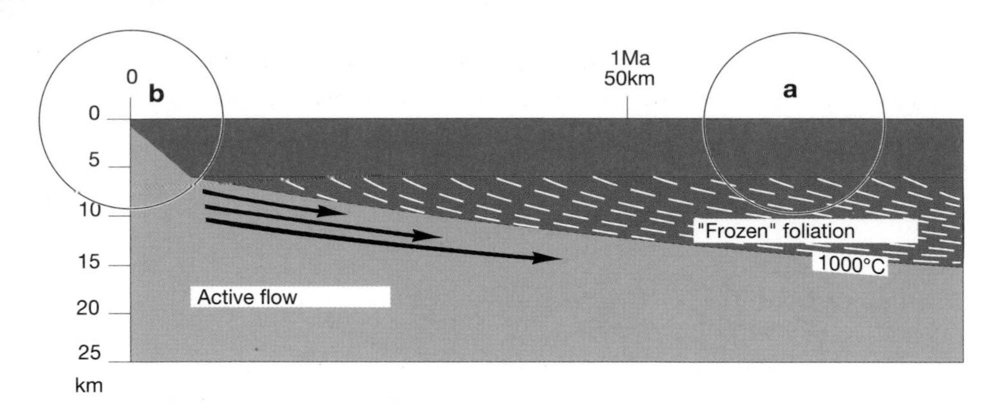

Figure 6.5 a,b
Indirect witness (**a**) or the culprit caught in the act (**b**). An ophiolite is a fragment of oceanic lithosphere usually sampled at some distance from the ridge (case **a**). However, sometimes it may sample the ridge itself, becoming solidified in the process (case **b**). The figure also shows how we may visualize the structure of plastic flow close to the ridge when we only have available the indirect evidence of a fragment of lithosphere sampled at some distance from the ridge. We know that the foliation of ophiolitic peridotites (*dashed*) represents flowing mantle being frozen on contact with the base of the lithosphere at a temperature of about 1000 °C. (A. Nicolas and J. F. Violette 1982, Tectonophysics, 81, 319–339)

gressively during sliding against the cooler lithosphere and it gradually slows down and ends up glued to this lithosphere. We then say that it accretes to the lithosphere. The lithosphere thickens in the process, and sliding of asthenosphere is shifted to increasingly lower levels as we move away from the ridge. The foliation planes in peridotites are derived from these flow planes in the asthenosphere which froze or became "fossilized" parallel to the lithosphere. We then conclude that in massifs of ophiolitic peridotites the foliations frozen just below the Moho represent sliding of asthenosphere in the immediate vicinity of the ridge and that the foliations recorded at depth reflect flow at some distance from the ridge, the distance growing with the depth of the foliation level below the Moho. From this we may conclude one rule for the analysis of structures in peridotites: the foliation planes in peridotites are usually parallel to the plane separating lithosphere from asthenosphere.

If all ophiolites formed like this, recording the asthenospheric movements in peridotites would always tell us how the assemblage moved away from the ridge yielding only indirect information on the movements which took place directly below the ridge. Fortunately, in some ophiolite massifs, the thrust during obduction may have scalped on active ridge cutting the mantle directly below the ridge

(Fig. 6.5b). The part of the mantle still attached to it cools under static conditions before or during the initial stages of this transport and furnishes us, through the intricate patterns shown in the map by the lineations and foliations, the trajectories of its ascent below the ridge.

The Roots of Volcanoes

The structural analysis of the peridotites sets us now on a new path, the melting of mantle and the extraction of basalt. We know already that asthenospheric mantle rising below a ridge undergoes partial melting during which basaltic liquid is drained from it. This liquid feeds the ridge and creates, on cooling, the 6 km of crust extending from the ridge.

Melting of the mantle and extraction of basalt, as described below are of fundamental importance for our understanding of the functioning and feeding of the oceanic spreading centres and for explaining the differences between continental and oceanic rifts and between slow- and fast spreading oceanic ridges (Chapt. 8). In a still more general way, this subject literally encompasses the sources of volcanism. Certain features of volcanic eruptions, such as their violence and their episodic nature, depend on it directly, as we shall see below.

Melting of Mantle

In order to understand melting of mantle below a ridge, we have to take pressure created by the weight of rocks into consideration. It is common knowledge that melting a solid or bringing a liquid to boil requires heat. A cook and a mountaineer also know about the effect of pressure on boiling. Increasing the pressure in a pressure cooker raises the temperature of the boiling point, thereby reducing the time required for cooking. In the thin air of mountain peaks the opposite effect takes place, and the temperature of the boiling point is reduced to below 100 °C so that at very high altitudes it is virtually impossible to hard-boil an egg.

In analogy to these observations, where mantle circulation below a ridge entails an upward component, the weight of the overlying rocks will decrease and the peridotites will undergo decompression. This is in the order of 300 bar for every kilometre of ascent, this value being the pressure at the base of a 1-km-high column of upper man-

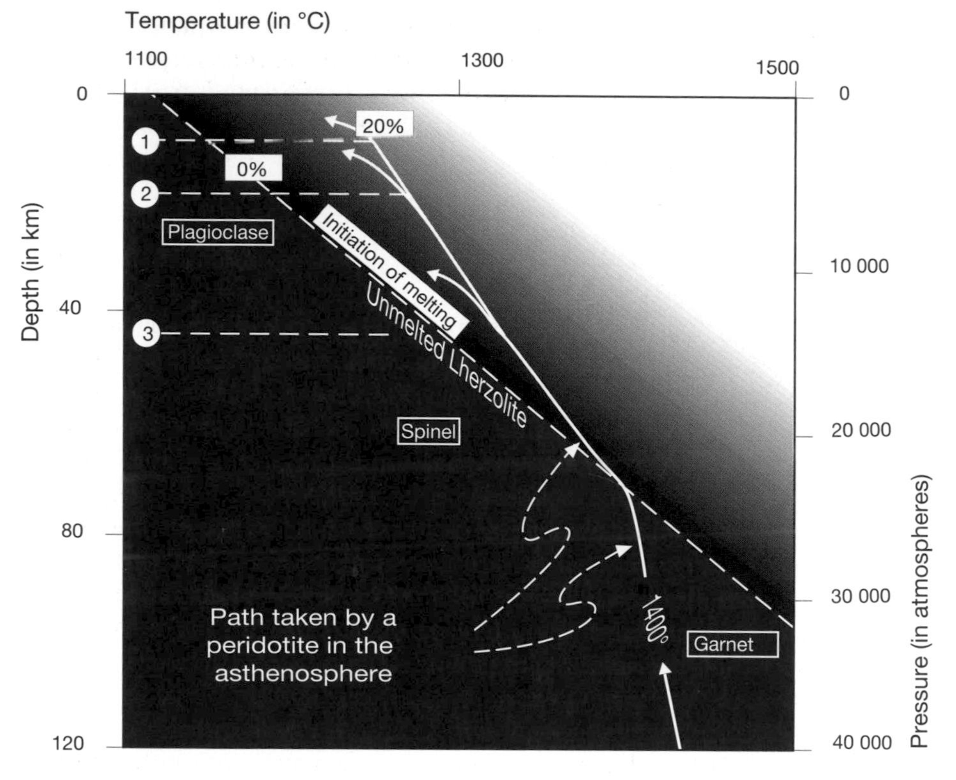

Figure 6.6
Diagram of the evolution of temperature and melting of a mantle peridotite during ascent to the surface. The trajectory follows the *white arrows from bottom to top*. Rising within the dry asthenosphere (*dark red*), the peridotite starts to melt when it crosses into the lighter-coloured areas. The increasing degree of melting is indicated by *increasingly lighter red colour*. The diagram contains two other pieces of information: In the "solid" domain (*dark red*), depending on the depth, the peridotite may contain garnet (below 75 km), spinel (30–75 km), or plagioclase (above 30 km). *1*, *2*, and *3* correspond respectively to three possible trajectories, depending on the depth at which the rising peridotite becomes incorporated in the lithosphere (cf. Chap. 8)

tle material with a density of 3.3 g/cm^3. As in the case of water boiling earlier at lower pressure, the decompression resulting from an upward movement leads to a lowering of the melting point. The average temperature conditions in the mantle are such that a peridotite will start melting when it rises above approximately 75 km. In Fig. 6.6, the curve marking the start of melting of peridotite as a function of depth, the so-called solidus, is shown together with the change of temperature in the mantle, the geothermal gradient. This latter curve indicates the temperature a peridotite will have at a certain depth. A peridotite rising in the mantle is represented in the diagram by a point following this curve to the top. The two curves intersect each other at about 1400 °C, i. e. at this depth a peridotite, the temperature of which followed the geothermal gradient during its ascent, encounters the melting curve and thus will start melting. The main uncertainty about this depth lies in the fact that the geothermal gradients in the mantle are still poorly known and that they may vary especially close to hotspots.

This description is only applicable as long as the peridotite remains in the asthenosphere in which, according to our restrictive definition, there will be no heat loss through conduction. Thermodynamically speaking, this system is adiabatic and melting in it is produced by "adiabatic decompression". As during its ascent our peridotite penetrates into the lithosphere, things are changing. In contact with the cooler peridotites of the lithosphere, it starts to lose heat by conduction. Its temperature drops, melting stops on the trajectories above points 1, 2 and 3 in Fig. 6.6, and it becomes frozen to and integrated in the lithosphere. In summing up, let us draw some conclusions highly relevant to our further studies:

- To get melting under way, it will suffice if part of the asthenosphere rises to a depth above 75 km. This may be in the form of an enormous slowly rising dome or a plume rising very rapidly from depth in a convective mantle. Anyhow, let us for the time being exclude special environments like hotspots and equate the asthenosphere to a lherzolite depleted of water. Water effectively lowers the melting point and, as a consequence, facilitates early melting. This comparison is quite acceptable as there is very little water in most of the mantle. This is no longer the case above a subduction zone where melting of mantle is induced by dehydration of the subducted crust. This explains why mantle below a lithosphere of less than 75 km thickness, and consequently still young, may be affected by notable melting. In contrast to this, mantle of older regions where the lithosphere has attained its normal thickness of about 100 km or more will usually remain dry.
- The molten fraction extracted from asthenospheric mantle becomes larger as the latter ascends further, a conclusion drawn from Fig. 6.6. We have seen already that the line representing asthenospheric peridotites intersects the melting curve of lherzolites near 75 km. When the peridotites continue to rise,they penetrate further into the domain of melting, i. e. the degree of partial melting will increase. Assuming that the peridotite was to contin-

ue its ascent to the surface without losing heat, ignoring the existence of lithosphere above the asthenosphere, it could produce approximately 25 % of basalt. This is quite clearly a highly theoretical situation, as anywhere on our planet a layer of lithosphere covers the asthenosphere, stopping the latter's continued ascent and melting. Over oceanic ridges lithosphere is thinnest, a few kilometres at the most (Fig. 1.9) and melting of mantle here is most pronounced with values surpassing 20 % (case 1 in Fig. 6.6). Under the Limagnes of the French Massif Central and in a more general fashion under the rifts where the lithosphere attains a thickness of some 50 km, there is also melting of the underlying asthenosphere but the melted fraction is notably smaller than under the ridges (case 3 in Fig. 6.6).

The Extraction of Basalt: Field Evidence

In addition to signs of high temperature deformation, peridotites frequently also preserve indications of melt circulation. Patches or veins of basaltic liquid are frequently "frozen" in the rock during cooling. This allows us to follow the history of its melting and to see the mechanism of basalt extraction in action. We are able to do so because we have seen that the extraction of basalt took place during the deformation accompanying the ascent of asthenosphere below the ridge and because we know that these deformations were important at very high temperatures. This deformation is thus not tied to some type of very late tectonic event overprinted to the peridotite during its cooling stages. In other words, the best evidence for asthenospheric deformation in a peridotite lies in the presence of traces of magma in the deformed structures (Fig. 6.7).

However rapidly the peridotite cooled after melting, the basaltic liquid has sufficient time to crystallize as gabbro, a rock rich in light-coloured plagioclase and because of this standing out from the dark background of the peridotites. These manifestations of melting are clearly recognizable in the field (Fig. 6.8). the melted liquid, initially dispersed as droplets throughout the peridotite, starts to assemble in these films during deformation (Fig. 6.8a). The films grow in size, start to join up, and form a continuous network of liquid within the rock like water filling a sponge. In the next and decisive stage the up to several tens of centimetres thick veinlets and dykes start to coalesce from the sponge and drain the basaltic liquid towards the surface (Fig. 6.8b, c). In their root zone the veins are irregular like the

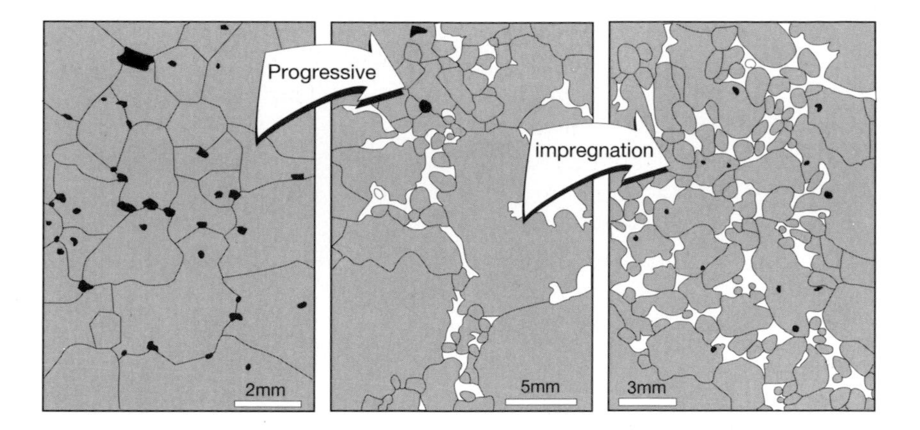

Figure 6.7
Drawings from thin sections illustrating the modifications in microstructure of a peridotite by progressive impregnation with basaltic liquid (*white*). *On the left* is a dry peridotite showing the clear outlines of olivine crystals and chromite grains (*black*). The basaltic liquid on the *right*, under increasing impregnation, was able to crack the peridotite and disperse olivine fragments in the liquid prior to the state of compaction preserved now. (A. Nicolas 1989, Structures of ophiolites and dynamic of oceanic lithosphere, Kluwer Academic Publ., 367 p.)

roots of a tree, gathering the small pockets of liquid present here and there (Fig. 6.8b). Higher up in the peridotite massifs, the veins become more rectilinear and cut right across the peridotites, being comparable in section to the trunk of a tree. This comparison, however, should not be strained. The dykes most probably do not take the shape of a cylinder or tube, but rather of sheets becoming increasingly flattened as the surface is approached. From these dykes the magma is then expelled to the top of the ridge or to the Earth's surface in regions of subaerial volcanism.

Figure 6.8 a–c
Successive stages in the extraction of basaltic liquid from peridotites. Now frozen as gabbro, the basaltic liquid appears *white* in the photos. **a** First stage of concentration of the liquid in films. **b** Extraction by veins gathering the liquid from adjacent films. **c** Injection of the basaltic liquid as a dyke into an already well-cooled peridotite

Melt Can Fracture the Peridotite

What force would be able to fracture rocks as resistant as peridotites, especially after they have cooled?; and to maintain these fractures open against the formidable pressure of the enclosing peridotites, allowing the potent basaltic liquid to circulate in them?; and eventually to produce the sumptuous outflows of lavas through the throats of volcanoes? This force is quite simply the Archimedan pressure law: "Any body placed into a liquid will experience a vertical pressure equal to the weight of the fluid displaced." In which way could we apply this knowledge to the extraction of basaltic liquid from the mantle?

Let us clarify that the mantle in the zone of partial melting is highly deformable and on a geological scale behaves like a fluid keeping in mind that at all instants peridotite remains solid. This

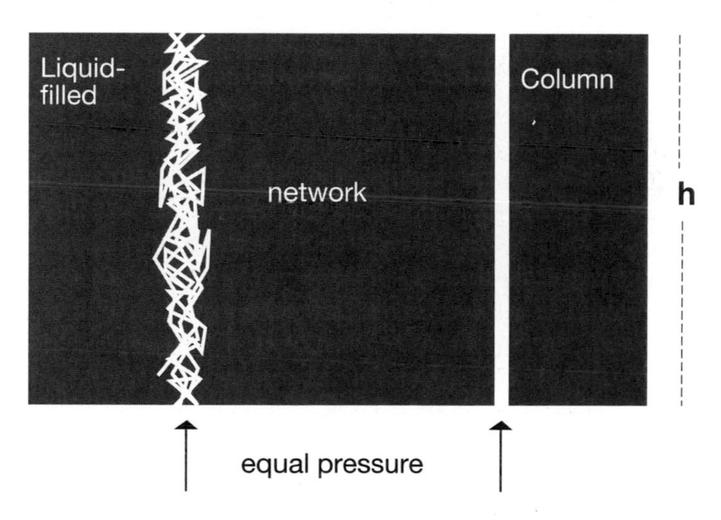

Figure 6.9
The Archimedian principle in peridotites. An irregular, but uninterrupted network of basaltic liquid is equivalent to a regular column of the same height

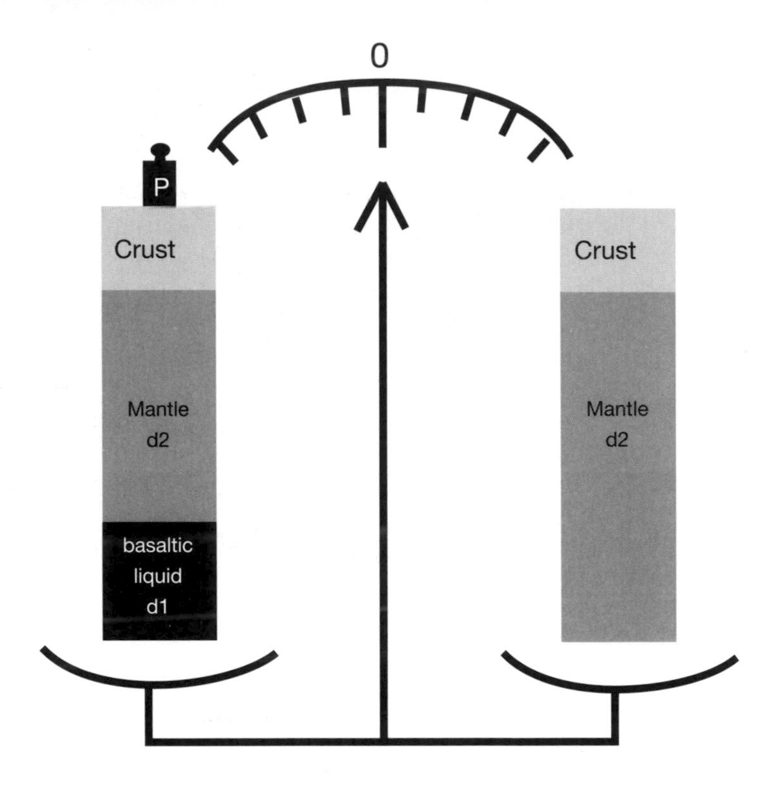

Figure 6.10
How to create an overpressure *P* above a column of liquid by applying the Archimedian principle

mantle is our Archimedan liquid whereas the zones of basaltic liquid represent the foreign body. The basaltic liquid, forming a continuous network of height h in the peridotite, may be compared to a column of the same height (Fig. 6.9).

The origin of this force in the zone of mantle melting may be illustrated in a simple manner by comparing on our scale for weighing lithosphere (Fig. 6.10) the weight of two columns of height z down from surface to the depth considered. The left column contains at its base liquid basalt of height h with a density d1 = 2.8 g/cm^3. The right column is identical in height but the liquid basalt portion is replaced by mantle material with a density d2 = 3.3 g/cm^3. Quite obviously, the column with basaltic liquid at its base would be lighter. We therefore would have to place a weight P on the corresponding side of our scale. When we have chosen the columns with identical surface areas, the weights will correspond to pressures as weight = pressure x area. The pressure dif-

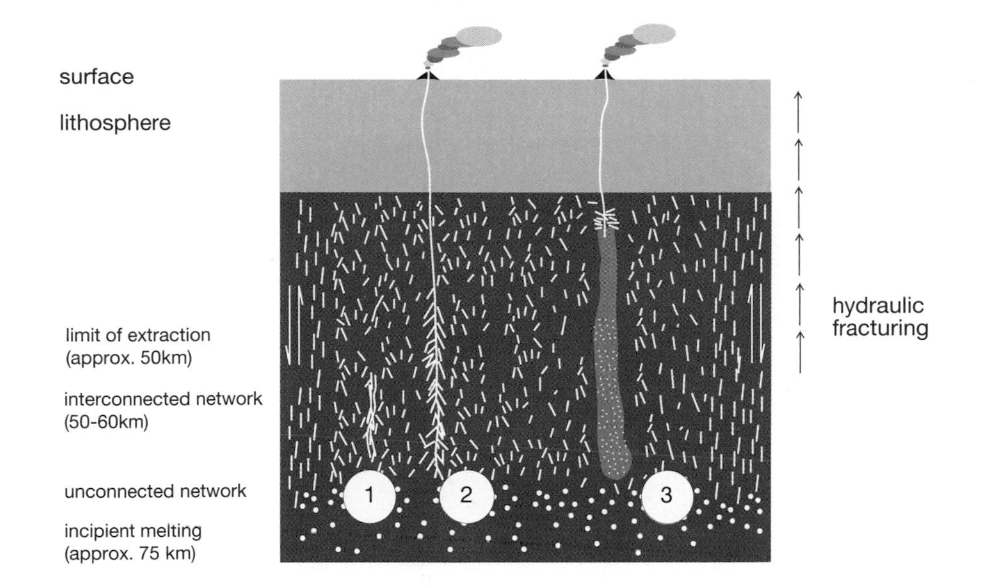

Figure 6.11
How the liquid may fracture rocks and form a volcano. At *stage 1*, a column in the sense of Fig. 6.9 is formed. At *stage 2* the column passes a critical height at which the fluid pressure at its top exceeds the resistance of the surrounding rocks and a fracture is formed which rapidly propagates to the surface, leading to a volcanic eruption. At *stage 3* the fracture closes up by draining the liquid from the adjoining peridotites, leaving in its wake depleted peridotites (harzburgites and dunites). (After A. Nicolas 1986, J. Petrol., 27, 999–1022)

ference between the two columns then is P = h (d2-d1) = 0.5 x h (gram, centimetre).

Expressed in bar and kilometres, this overpressure at the top of a 1-km-high column of liquid will amount to 50 bar against the surrounding peridotite. This overpressure results from the buoyancy of the liquid which, being lighter, will attempt to rise. Assuming that the column attains a height of 10 km, the corresponding pressure would be 500 bar, a pressure that would exceed the resistance of the surrounding peridotite and fracture it. As a consequence, a fissure would open and basalt would flow into it, propelled by the overpressure. This phenomenon, called hydraulic fracturing, is well known from civil engineering and oil exploitation. As the fissure advances upwards, the height of the column of liquid in Fig. 6.11 increases. The overpressure increases as well, and theoretically the process is accelerated, eventually culminating in a cataclysmic volcanic eruption. Fortunately, the resistance exerted by the

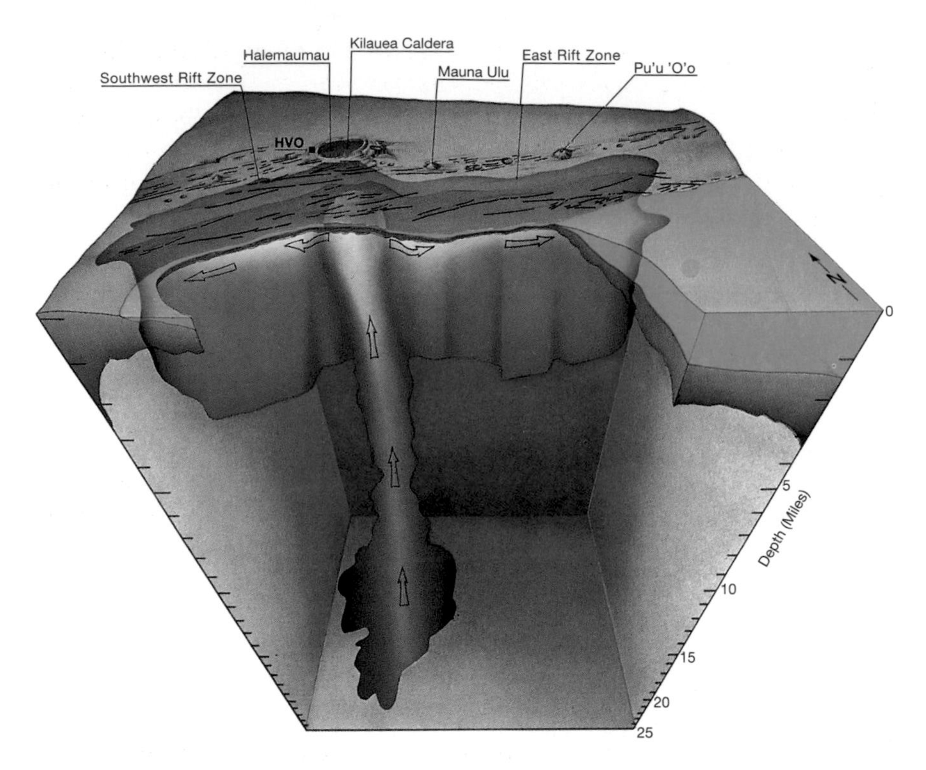

Figure 6.12
Contours of the zone of magma extraction below the volcanoes of the Island of Hawaii. (M. P. Ryan 1988, J. Geophys. Res., 86, 7111–7129)

rocks to the propagation of the fracture and the friction of the liquid moving in the fractures harmess the energy in question. Under volcanoes, at least during the active phase, there is thus an irregular column of peridotites fractured by moving basaltic liquid. By detailed analysis of earth tremors resulting from this fracturing, it was possible to reconstruct the shape of this column under the volcanic island of Hawaii (Fig. 6.12).

When connection with the surface has been established and the lavas flow out on the flanks of the volcano, the supply of basaltic liquid contained at depth in the fracture system becomes sucked up by the same Archimedan force (case 3 in Fig. 6.11). The fractures close up and the column of fractured peridotite dries out. A new eruption will start when in the root zone of the dykes a network of basaltic liquid becomes interconnected again to a critical height, h, which has been estimated above to be about 10 km (cases 1 and 2 in Fig. 6.11).

Episodic Nature and Violence of Volcanism

The extraction of basaltic magma from the mantle through fractures and dykes is thus a violent phenomenon. The fracture opening the way for the magma advances very rapidly, producing seismic tremors on its way. When observed under volcanoes, these tremors will herald a new ascent of magma. In the mantle, the magma itself circulates in the fracture or dyke with velocities in the range of 1 km/h. Closer to surface, the velocity can increase as a result of the buoyancy exerted by gases liberated from the liquid because of the drop in pressure. It will take only a few weeks for a dyke to form 1 m of new oceanic crust over the 6-km thickness of the crust under the ridge. This figure of 1 m is not chosen arbitrarily. It is the average thickness achieved by the decimetre-thick veins draining basalt from the mantle by the time they have reached the dyke complex at the top of the crust (Fig. 5.1). Oceanic crust is thus formed in a discontinuous fashion each time a column of magmatic liquid opens up to drain the mantle (Fig. 7.8). The crust spreads approximately 1 m over a period of a few weeks, and then settles for a few years or decades. These predictions from ophiolitic studies were verified in 1993 along the Juan de Fuca Ridge extending west of Oregon, where a submarine basaltic eruption was observed in June 1993. The seismic monitoring of the feeding dyke system showed that activity lasted only 2 weeks. The period of quiescence depends on the spreading rate; if it is 10 cm/a, it will take 10 years on average to prepare the crust by elastic stretching for the 1 m of opening corresponding to one dyke. At a rate of 1 cm/a it will take 100 years.

In this short outline of the origin of volcanism, we have discovered how the ascent of asthenosphere and the expansion of a ridge, phenomena which normally proceed at a geological rate of a few centimetres per year, may induce violent instabilities. Indeed, the rates of lava ascent may be 10 million times higher than those of the peridotites from which they are derived. They may be even higher still when we take into account the degassing of the lavas close to surface in subaerial volcanoes. It is believed that analysis of magma extraction from the mantle, shown here in a ridge environment, may also be applied to other environments like intracontinental volcanism, hotspots, island arcs etc. Certain basic similarities suggest that the physical aspects of the phenomenon are identical.

7 The Forges of Vulcan in the Kingdom of Neptune

It all starts with the diapirs, these thin plumes in the mantle which, being lighter because of their content in basaltic liquid, rise and spread out under the ridges, allowing the basaltic liquid to escape from them periodically. This basalt also impregnates mantle peridotites located at the top of a diapir to the extent of dispersing them within a magma at the level of the Moho below a ridge. Higher up, it feeds a magmatic chamber with domed walls, breaks through the roof of this chamber, intrudes parallel to the ridge in a seam of liquid, the dyke-to-be in the dyke complex. Finally, it feeds submarine basalt flows. Thus the ridge system, confined within only a few kilometres on either side of the axis, is a result of the energy of a convection cell of the mantle with dimensions tens to hundreds of kilometres deep. The spacing of these diapirs in the mantle below the ridge controls the segmentation of the latter.

Continuous Flow

How does a ridge function, this factory producing oceanic lithosphere? By which secret processes will the mantle which rises below a ridge, hot but still entirely solid at a depth of at least 100 km, start to split higher up into a sterile residue and a crust made up of chemically varied rocks? How can we obtain in this crust a layering as fine as in the best-developed sedimentary rocks (Fig. 5.1) or dykes intruded in such a regular pattern into each other? How could we explain that in the gigantic circulation of material pervading the terrestrial mantle, a mantle dome extending over the area of a state may focus its effects over a few kilometres on either side of the ridge axis as we have seen (Fig. 3.4)? Why should the lavas extrude and the faults cut down only within this very narrow band? Is it not extraordinary that this mechanism operates at such a constant rate creating a crust with a uniform thickness of 6 km and the uniform stratification revealed by seismic soundings and the observations in ophiolites? This regularity contrasts markedly with the extraordinary complexity of the continental crust.

It is, however, necessary to modify to some extent this remark on the regularity of the processes encountered on ridges. The longitudinal segmentation which cuts the ridges between two transform faults or two overlapping spreading centres (Chap. 3) suggests that the process of formation of lithosphere does not operate uniformly along strike. We shall return to this question later. Furthermore, when talking about the slow ridges in Chapter 8, we shall see that the formation of lithosphere is probably not a continuous process. A slow ridge functions as a blast furnace supplying molten flows in a discontinuous fashion. In fast-spreading ridges, we are dealing with a continuous flow as in certain foundries, incessantly producing new lithosphere. This new lithosphere is identical throughout time as sampling here and there in ophiolites demonstrates. Studying a fast-spreading ridge in which the properties do not change with time or only very slowly will be more simple than studying a slow ridge which passes between phases of activity and quiescense. Most ophiolite belts, and especially the Oman one, are comparable to fast-spreading ridges (Chap. 5). We shall now study the functioning of a fast-spreading ridge based on data from the East Pacific Rise and the Oman ophiolites. In Chapter 8, we shall then try to find out in which way the functioning of slow ridges and rifts differs from this situation. A large part of the data and concepts presented in this chapter has been gleaned from our studies in Oman.

A Well-Combed Mantle

Maps of asthenospheric movements in the vicinity of ridges obtained from systematically recording foliations and lineations in peridotites of ophiolite massifs (Fig. 7.1) revealed an image confirming the situation expected: "well-combed" asthenosphere sliding parallel to a little-inclined plane separating it from the lithosphere. The asthenosphere thus carries on its back the newly created lithosphere, moving with it away from the ridge approximately at right angles (Fig. 7.2).

Seismic studies in the vicinity of the East Pacific Rise confirm this transport direction. They are based on a comparison of the propagation velocities in different directions over the ridge. We know from the olivine crystal that the waves propagate in the direction of slip (Fig. 7.2) faster than in any other direction. We say that olivine as well as the peridotite, which is composed of aligned olivine crystals, possesses a seismic anisotropy. Seismic soundings around the ridge show that the wave velocities are highest at right angles to the ridge and, inversely, slowest parallel to the ridge. From this we may conclude that in mantle peridotites, sliding takes place at right angles to the ridge as shown schematically in Fig. 7.2. We note also that this lithospheric drift approximately at right angles to the ridge may be deduced, on a large scale, from a map of the plate movements (Fig. 3.5). From the East Pacific Rise, the Nazca Plate drifts to the east at 5.6 cm/a and the Pacific Plate to the WNW at a rate in the order of 10 cm/a.

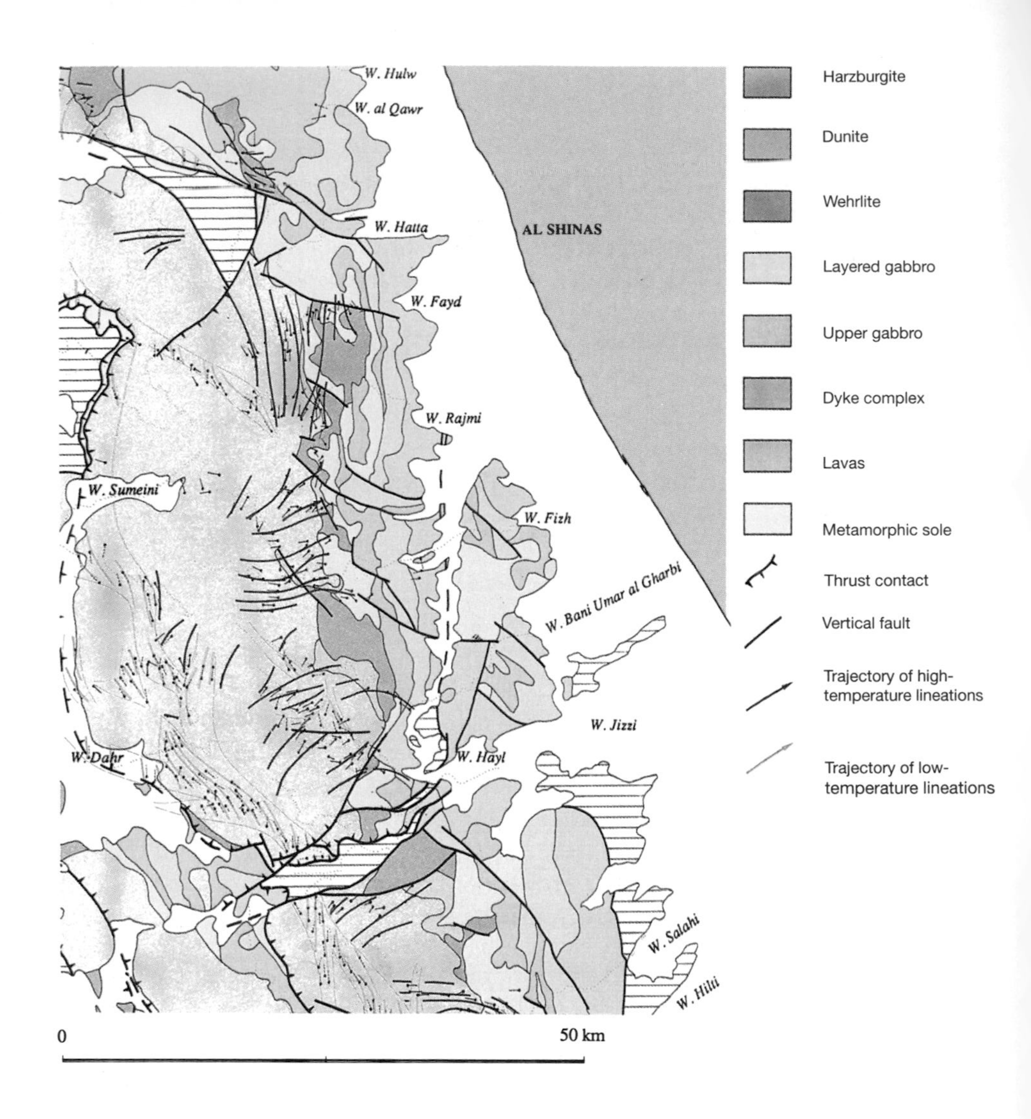

Figure 7.1
Structural map lineations in peridotites in parts of the Oman ophiolite belt. As the massif lies nearly horizontal, we may assume that the lineations and the trajectories reconstructed from them imprinted themselves onto the same foliation surface. The red trajectories which reflect flow of the very hot mantle away from the ridge axis are in a large part of the map perpendicular to the dyke complex which runs approximately N-S and is itself parallel to the ridge. (After E. Ball et al. 1988, Tectonophysics 151, 27–56)

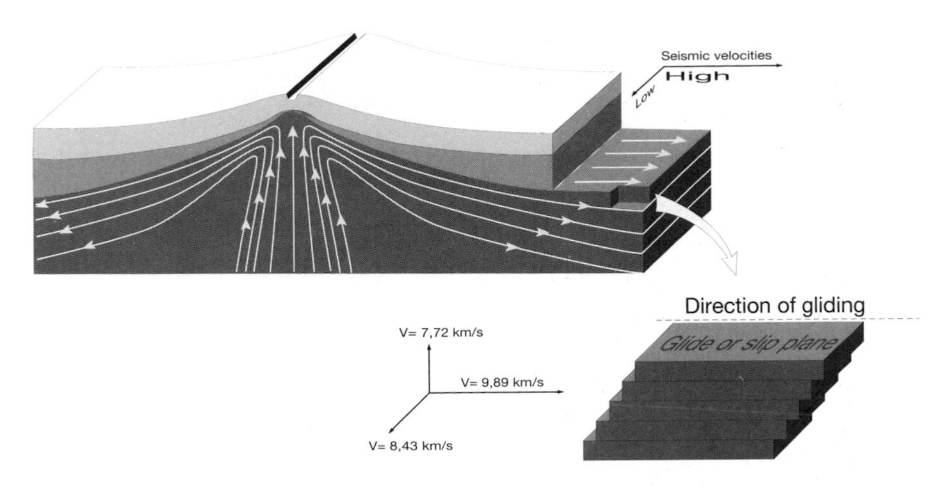

Figure 7.2
Relationship between large-scale motions of hot mantle away from the ridge and anisotropy of seismic velocities. The *inset* shows that gliding within the olivine crystals which assures the movement of the entire mass, oriented the crystals in such a way that the largest seismic velocity is parallel to the direction of the movement. This relationship is of special interest as it permits us to predict active or fossil flow directions by measuring the seismic velocities in the upper mantle

Diapirs – Mushrooms Under the Ridge

Unfortunately for simplistic models, mapping in certain ophiolite massifs like Newfoundland has also revealed that on the near-horizontal asthenospheric foliation planes there may be directions of movement oblique or even parallel to the axis of the presumed ridge as defined by the orientation of the dyke complex. The situation became even more complicated with the discovery in the ophiolite massifs of Cyprus and the Philippines and then in a much clearer way in several localities in the Oman belt, of highly inclined foliations which in the maps of the areas concerned define a closed trajectory extending at depth as an irregular cylinder several kilometres in diameter. The associated lineations are also steeply dipping and follow on average the axis of the cylinder (Fig. 7.3). Thus, the fossil traces of asthenospheric diapirs under ridges were first discovered in ophiolites. These structures recall the well-known salt domes or diapirs of northern Europe.

The word diapir refers to mushroom structures (Fig. 7.4) produced by the salt rising through its sedimentary cover. This rise is

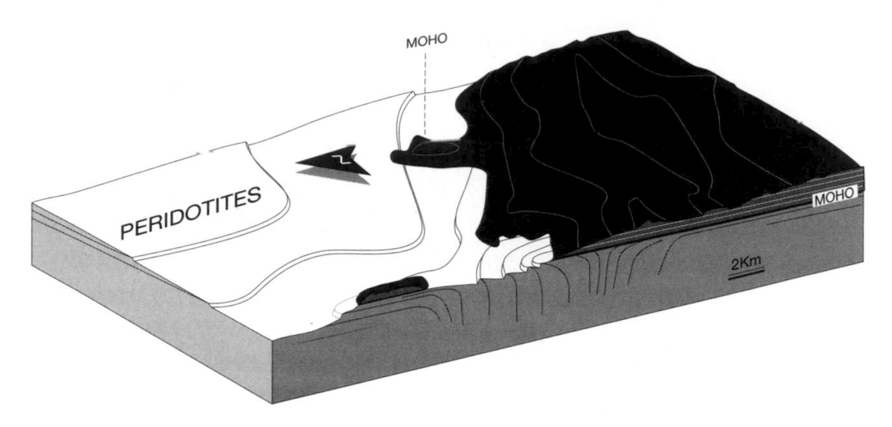

Figure 7.3
Block diagram of the Maqsad area of Oman showing an outcropping mantle diapir still partly covered by layered gabbros (*black*) of the crust. Note the sharp rotations of foliation (*thin lines*) in the mantle just below the Moho. (After G. Ceuleneer et al. 1988, Tectonophysics, 151, 1–26)

caused by the fact that the density of the salt is lower than that of the sedimentary cover, the latter pressing the salt upwards. The process is assisted by the pronounced plasticity of salt on a geological time scale. We may compare the diapirs also to plumes of smoke which, being lighter than the ambient air, rise from a chimney. Applying this conclusion to mantle diapirs supposes that the ophiolites may be interpreted as samples of an active ridge, as schematically shown in Fig. 6.5b. Detailed field analyses have shown that the grossly cylindrical foliations represent a chimney channeling a vertical flux of asthenosphere to below the Moho where it then spreads out in all directions below the ridge. This explains the older observations of asthenospheric movement oblique to the axis of the ridge as noted in the Newfoundland ophiolites. These massifs thus must have been derived from a zone of oblique divergence around a diapir. Mapping on a larger scale as in Oman permits us now to localize such zones with respect to the centre of a diapir (Fig. 7.3).

In the immediate vicinity of the ridge the divergence of mantle flux around diapirs is radial. How can we explain that from a certain distance onwards the flux tends to turn perpendicular to the strike of the ridge? An answer to this question may be provided by the role of the fracture zones (Fig. 7.6). Because of the displacement between the two sides (Fig. 3.6), a fracture zone can place older and thus thicker lithosphere against younger and thinner lithosphere or even

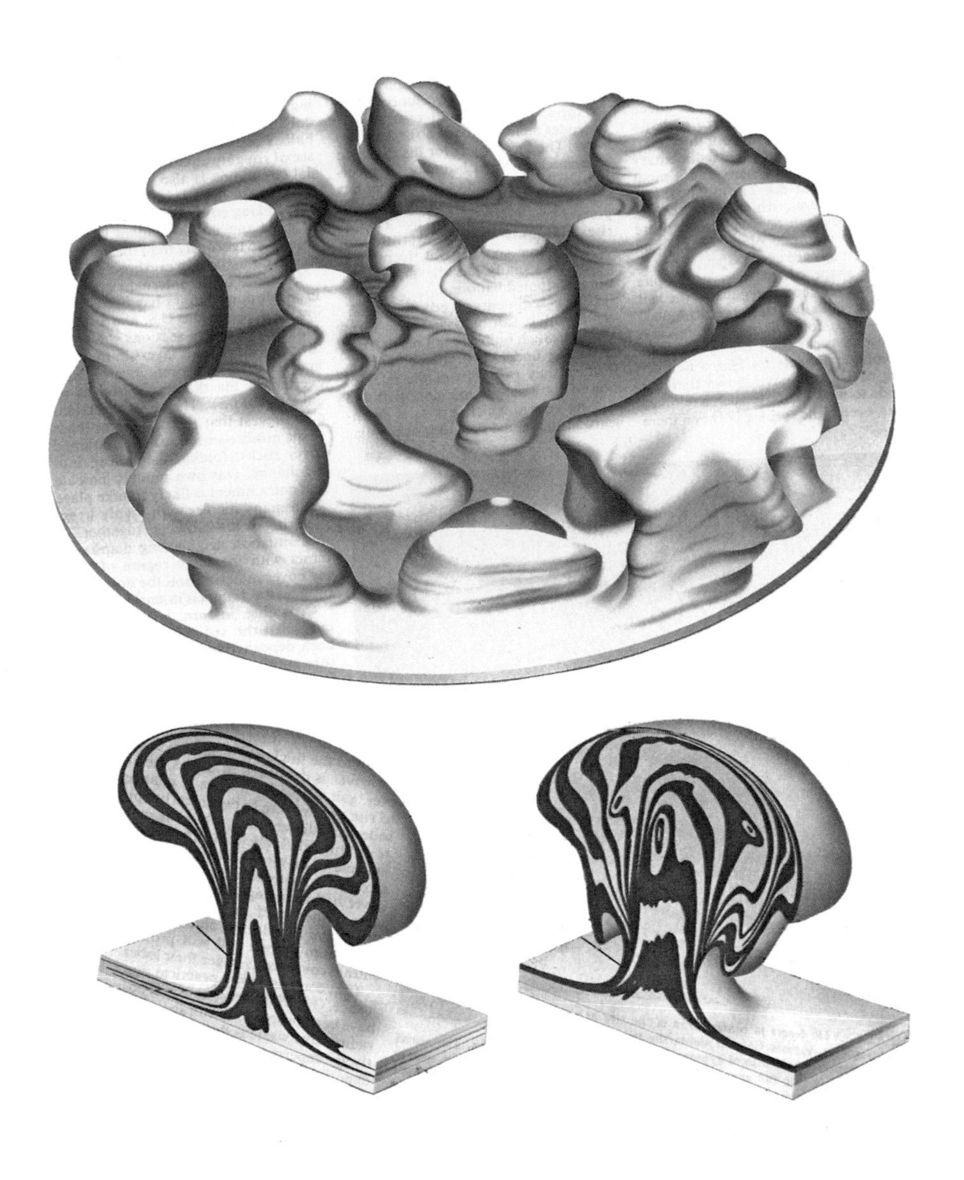

Figure 7.4
Experiments of intrusion (diapirism) carried out in centrifuges with initially flat-lying layers of plasticine of equal thickness. (C. Talbot and M. Jackson 1987, Pour la Science, 120, 46–57)

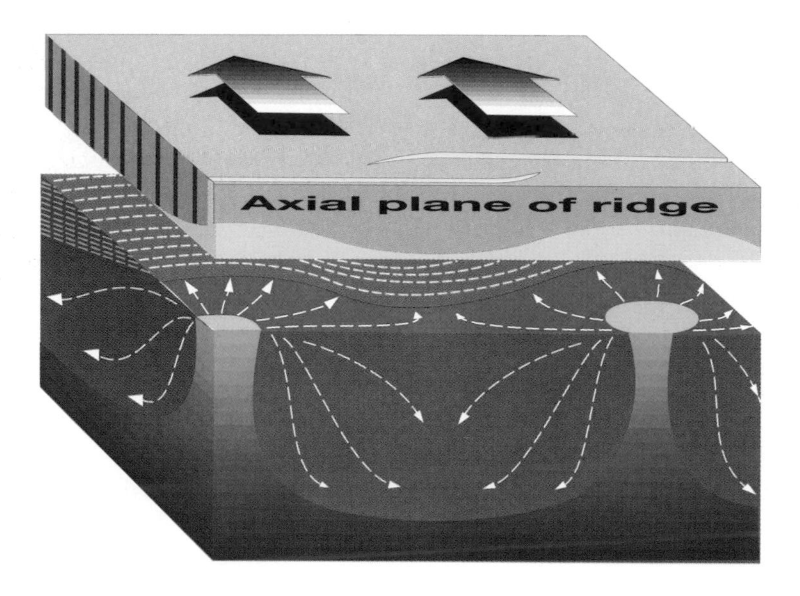

Figure 7.5
Diapiric rise of mantle material below a ridge and its envisaged relationship with the segmentation of the ridge. The partly molten mantle layer shown in *dark red* at the base of the figure generates the diapirs (ligher-coloured columns), the feeding channels for basaltic melt and material for the ridge. On the way up, the basaltic melt is squeezed out of rising peridotites. At the top of the diapir, they diverge (*white arrows*) and end up by accreting to the lithosphere (in *dark brown with white dashed lines* showing the foliation). Above it, the magma chambers of the crust (in *yellow*) underlie the segments which are separated on surface by overlapping zones. (A. Nicolas 1989, Structures of ophiolites and dynamics of oceanic lithosphere, Kluwer Publ., 367 p.)

against the ridge itself. The flux of young asthenosphere diverging from the ridge, once it encounters the cool wall of older lithosphere which it cannot penetrate, becomes channeled parallel to this lithospheric wall, and thereby at right angles to the ridge (Fig. 7.6). Close to the ridge this channeling effect will be notable only locally in the immediate vicinity of the fault but as we move away from the ridge, flow of asthenosphere parallel to the faults becomes increasingly important.

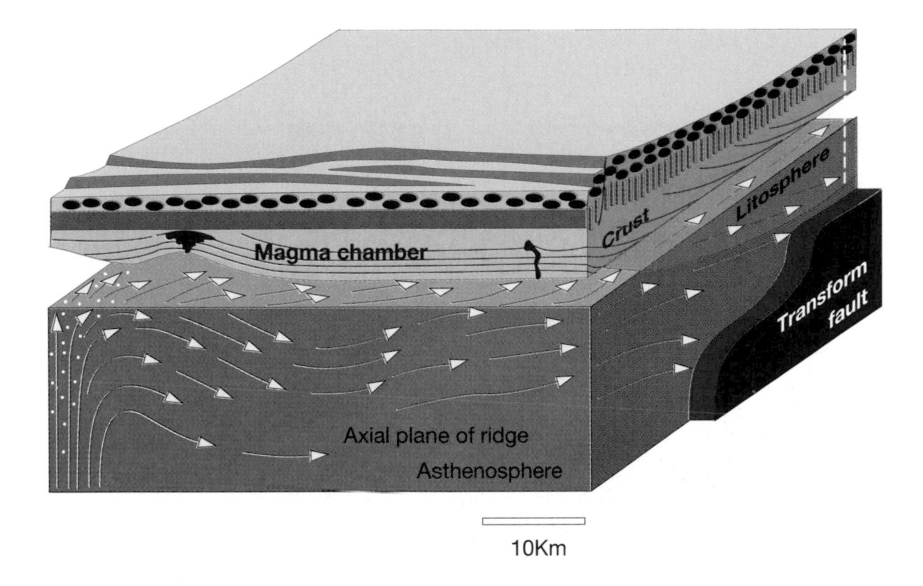

Figure 7.6
Flux of hot mantle (asthenosphere) rising in a diapir (*left*) and spreading from it into all directions and eventually parallel to the lithospheric walls of fracture zones. (A. Nicolas and J. F. Violette 1982, Tectonophysics, 81, 319–339)

Chimneys in the Mantle

From a depth of about 75 km onwards, athenosphere rising below a ridge will start to melt, the degree of melting increasing further during continued ascent (Chapt. 6). When arriving within 50 km below the ridge, the peridotites will contain 5–10 % of liquid. As the ascent continues, the melt portion will not increase further, and from 5–10 % onwards, the liquid will cease to be dispersed in droplets and scattered films and will start to form an interconnected network extending over increasing distances, thereby facilitating the extraction of melt from the environment of origin and its expulsion towards the surface. A side-effect of this extraction will be that the fraction of liquid stable in the meltling peridotite stabilizes around the level at which the liquid starts to form an interconnected network.

The presence of basaltic liquid in the peridotite decreases its density and improves its plasticity or, correspondingly, lowers its viscosity. Although being only modest – 5 % liquid in the peridotite will lower its density by 1 % – these changes may entail considerable effects as they affect an already highly deformable material over geological peri-

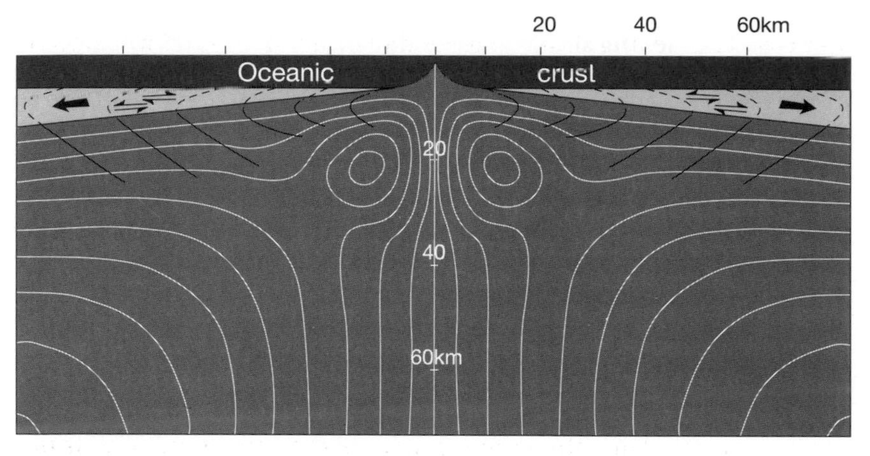

Figure 7.7
Computer-generated model of mantle diapirism below a ridge. The increasing drop in density of the mantle resulting from the presence of basaltic melt forces a small diapir under the ridge which is then expelled laterally (*black arrow:* forced flow deduced from studies in ophiolites). The local diapiric rise is superimposed onto the regional pattern of mantle convections (diverging flow lines at the *bottom* of figure). (M. Rabinowicz et al. 1984, Earth Planet. Sci. Lett., 67, 97–108)

ods. This may be studied by numerical modelling in large computers or, in a more perceptible manner, in small so-called analogue models in which the mantle is simulated by layers of oil or honey. These studies revealed a situation similar to salt diapirism as modelled in Fig. 7.4. The dome of partially molten mantle, which extends over large areas under the ridge and rises slowly, will become unstable and tries to escape towards the top more rapidly. These mantle chimneys or diapirs will detach themselves from this layer here and there, and especially under the ridge axis (Fig. 7.5), but probably also away from the axis, as suggested by the reduced model in Fig. 7.4a.

In Oman, these diapirs rose just to the Moho where the upward flow of the hot and still magma-laden peridotites came to a halt and started to spread laterally (Fig. 7.3). On the map their shape is very irregular and their size of some 10 km is rather modest. The distance between adjoining diapirs amounts to 50 km. Numerical modelling of diapirism enables us to construct a section at a right angle across a ridge (Fig. 7.7).

The Focussing Effect of a Ridge

Numerical modelling shows that the rate of ascent in a diapir is about twice the expansion rate of the overlying ridge. Because of this high velocity, the main component of the upward flow of hot mantle is contained in the central part of the diapir and is thus focussed to an area of less than 10 km on either side of the ridge axis. This is a remarkable focussing effect considering that the underlying dome in the mantle, from which the diapir is derived, is probably 20 times larger than the diapir itself. We have seen in Chapter 6 that it is this ascent which triggers melting in the asthenosphere. It has been concluded that the bulk of melting and extraction of liquid basalt from the mantle under a ridge will take place within the diapir. After extraction of this liquid, the peridotites arriving at the Moho are particularly sterile. Thus the convergence of the solid flow of mantle towards the ridge caused by the diapir leads a comparable convergence in the supply of basaltic liquid. The bulk of the basalt extract-

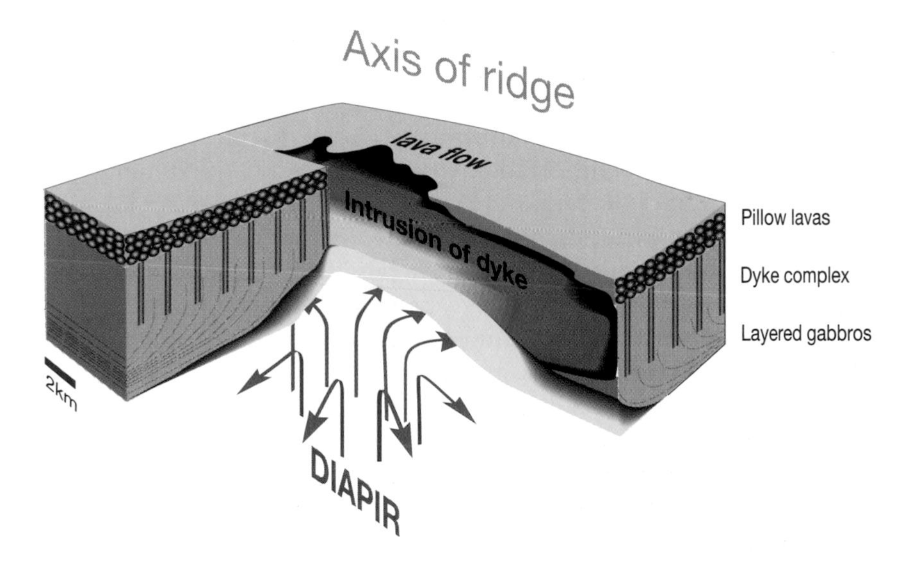

Figure 7.8
Model of ridge functioning above a mantle diapir. Fed directly by basaltic liquid tapped from the diapir, the chamber extends laterally over a few tens of kilometres. The figure shows the stage at which the ascent of fresh liquid opens up a new dyke in the dyke complex and feeds a submarine lava flow. (A. Nicolas 1989, Structures of ophiolites and dynamics of oceanic lithosphere, Kluwer Publ., 367 p.)

ed from the mantle transgresses the Moho under the ridge within a strip barely exceeding about 10 km in width. This conclusion is confirmed by observations on ophiolites which show that basaltic dykes only very rarely cut the gabbros, emanating from the magma chamber. This implies that very little basaltic liquid will rise in the crust outside the limits of the magma chambers. We also note that the magma chamber itself acts like a chimney, focussing the supply of basaltic liquid towards its roof, which has an area not extending more than a few kilometres in width (Fig. 7.8).

Is There a Moho Below a Ridge?

This question relates to the problem of how the abrupt divergence of flow in the mantle immediately below a ridge takes place and to the nature of the floor of the magma chamber. It has long been assumed that the floor was a pronounced horizontal plane upon which the gabbros are deposited from the base of the crust, upwards in successive horizontal layers like sediments accumulating in a body of water. However, the latest seismic data available on the situation below ridges, as well as field observations to be presented below, lead us away from this model and make us ask the question forming the title of this section.

In ophiolites the transition zone between crust and mantle (Fig. 5.8) is mostly very thin when taking into account the dimensions of a ridge system. Compared to the 6 km of crust overlying the more than 50 km of mantle directly involved in the activity of the ridge, the transition zone barely reaches 500 m of thickness above the diapiric zones in Oman and becomes reduced to only a few metres away from these zones. But how can it be defined? Lithologically the roof of the transition zone is made up of the layered gabbros of the base of the crust. The zone itself consists predominantly of dunites, the peridotites "dried out" by the extraction of magma. They contain only olivine and chromite, a spinel enriched in chromium because of the removal of the other elements by the basaltic liquid. The chromite may accumulate to form the chromite deposits exploited in a number of ophiolites. Towards the base, the transition is rooted in the harzburgites of the mantle. In the vicinity of diapirs, the transition zone is pervaded by gabbro dykes which frequently feed thick horizontal sills. Their contacts with the dunite may become diffuse and they may disappear in large cloudy masses.

Thus, the transition zone locally still harbours traces of intense impregnation by basaltic liquid, which has risen through dykes from the zone of extraction some 40–50 km lower down. Traces of this liquid passing through the harzburgites below the transition zone are present in the form of gabbroic dykes surrounded by metre-scale sheaths of dunite. The latter formed through the reaction of the liquid with the surrounding harzburgite. After the basalt has flowed out on surface and the dyke has become empty, it is frequently only this sheath of dunite which remains as a trace of the passage of the basaltic liquid (Fig. 7.9).

The dunites of the transition zone formed in the same way. To explain their extent of a few hundred metres, we have to assume that the basaltic liquid invaded the transition zone and was not constrained to the dykes. The vast number of dykes and the locally important impregnation in the dunites are evidence for this intense melt invasion. The transition zone under the ridge thus appears like a sponge of mantle material filled with basaltic liquid in contrast to the underlying mantle, where this liquid circulates by making use of fractures which are open over short periods of time only.

The transition zone deserves its name also as far as structure and movements of material are concerned. Outcrop conditions in ophiolites permitting, we may observe that the foliation planes in the peridotites of the transition zone are parallel to the Moho separating these peridotites from the also parallel layered gabbros of the crust. In Oman, moreover, the lineations in the peridotites are parallel to those in the lowermost gabbros. The deformation suffered by peridotites and layered gabbros is evidence for sliding at very high temperatures over considerable distances. There is thus a perfect structural continuity between peridotites and layered gabbros on either side of the Moho. However, detailed structural studies reveal a fundamental difference between these two rock types. The peridotites have undergone this deformation in the solid state, even if there was liquid trapped between the crystals whereas the layered gabbros were deformed in a magmatic state, i. e. like a mush of crystals suspended in the basaltic liquid. In the latter case, the flowing basaltic liquid carries along and orients the crystals just as a river carries along and orients tree trunks floating on its surface. The transition between deformation in the solid state and the liquid state in Oman is located exactly at the Moho. This is, however, not a general rule, as in the Newfoundland ophiolites this limit rises into the layered gabbros to a few hundred metres above the Moho.

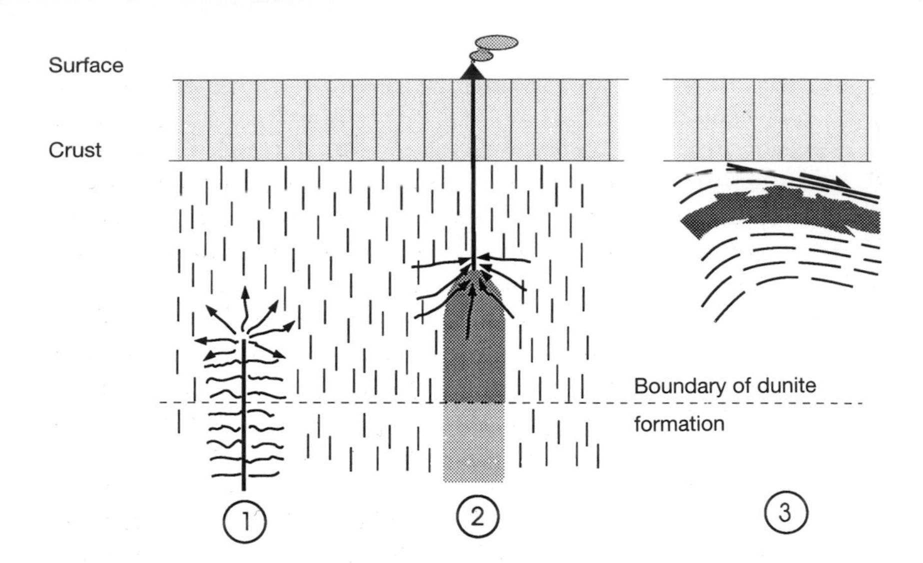

Figure 7.9
Origin of dunite veins frequently observed in ophiolitic peridotites. **1** A fracture filled with basaltic liquid seeks its way to the surface, invading the peridotite over several metres. Being filled with liquid, it reacts with the latter, dissolving its orthopyroxenes. **2** Having reached the surface, the fracture drains back the liquid injected during stage 1. It leaves in its wake peridotites drained of their liquid and transformed into dunites in the higher portions of the mantle where orthopyroxene can dissolve. **3** Lateral expulsion of the diapir deforms and "combs out" the dunite veins into the foliation of the peridotites. (A. Nicolas 1986, J. Petrol., 27, 999–1022)

Within a diapir, the rotation of mantle flow from vertical to horizontal (Fig. 7.3) takes place entirely within the transition zone over a thickness not greatly exceeding 500 m. On the scale of the diapir, such an abrupt rotation appears like a break in the rising flow. This also follows from "metallurgical" study of the structures in the peridotites of the transition zone and from a fluid mechanics analysis on the behaviour of such a system. In both cases, we have to conclude that there is a change in the physical state of the peridotites in the transition zone. A bent rod deforms into a regular curve up to the moment of rupture which results from an abrupt loss of resistance of the rod. In the same way, the quasi-rupture of asthenospheric flow in the diapir calls for to a similar abrupt loss of resistance, which is physically expressed by a sudden drop in viscosity by several orders of magnitude. The structures in the more or less impregnated dunites making up the transition zone furnish the explanation for this situation. In this zone there are still portions in which we may estimate from the fraction of gabbro present in the dunite (Fig. 6.7) that the fraction of liquid in the dunites below the ridge was so high that they ceased to react like a rock and disintegrated into crystals and rock fragments floating in the basaltic liquid. We know that this dispersion will start when the liquid fraction increases to above 35–40 %. Rotation of mantle flow would take place in this softened medium.

As a concluding remark, let us now revert to the heading of this section. The zone of transition below a ridge appears like a mushy floor to a magma chamber. There is a remarkable physical continuity between the mush of the peridotites in the transition zone and the basaltic magma on the road to crystallization in the magma chamber. It is therefore not surprising that this horizon is difficult to detect by seismic waves below the ridge and that the waves start to become reflected from the Moho only from about 5 km on either side of the axis onwards. This is the distance beyond which the "mantle sponge" has been drained by the pressure of the diapir.

The Origin of a Chromite Deposit – a Freak of Nature

Mineral deposits represent extraordinary concentrations of certain elements of minerals. Starting from a concentration of only 0.2 % in the peridotites, chromium, which we shall discuss now, becomes concentrated into dykes, pods and layers containing several hundred thousands of tons of this element, the largest deposits easily exceeding several million tons. Being several hundreds of metres long and several metres thick, the deposits are sometimes made up of pure chromite, a mineral of the spinel family containing up to 55 % chromium. Which peculiar process of nature might be able to concentrate the mean level of an element by a factor of about 300? For other metals even higher rates are known. For instance, the copper sulphides deposited at the foot of the black smokers on ridge basalts (Chap. 4) represent a concentration rate of 400 against the copper content of the basalts from which the metal has been leached.

Minable ore deposits are really freaks of nature created by the combination of exceptional factors. Understanding their origin will help in finding and exploiting them. In the case of chromitites associated with ophiolites we shall recognize how Nature operates.

We have seen that the basaltic liquid derived from partial melting of peridotite rises by upward opening and filling of a fissure which, upon cooling, forms a dyke. The first minerals start to appear in the basaltic liquid when cooling sets in closer to the surface. In most ophiolites, crystallization starts at 1200 °C within the magma chamber below the ridge and the first minerals crystallizing from the basaltic liquid will be chromite and olivine. The chromite, which usually precedes the olivine, fixes most of the chrome contained in the basaltic liquid. Crystallization continues in the magma chamber where these two minerals become diluted by the crystallization of plagioclase and clinopyroxene, the main constituents of the gabbros. However, crystallization may also commence within a basaltic conduit in, for example, the transition zone which, when spreading out below a ridge, starts to cool, its temperature dropping to below 1200 °C. Usually the chromite and olivine which appeared in this conduit are taken up by the flow of liquid and tend to become dispersed in the crust. It may, however, happen that the conduit opens up locally from a few centimetres to several metres in width which represents the size of the future chromitite dyke (Fig. 7.10a). This widening represents a trap for chromite and olivine. The crystals introduced by the liquid and those which formed along the cooler walls of the cavity are vigorous-

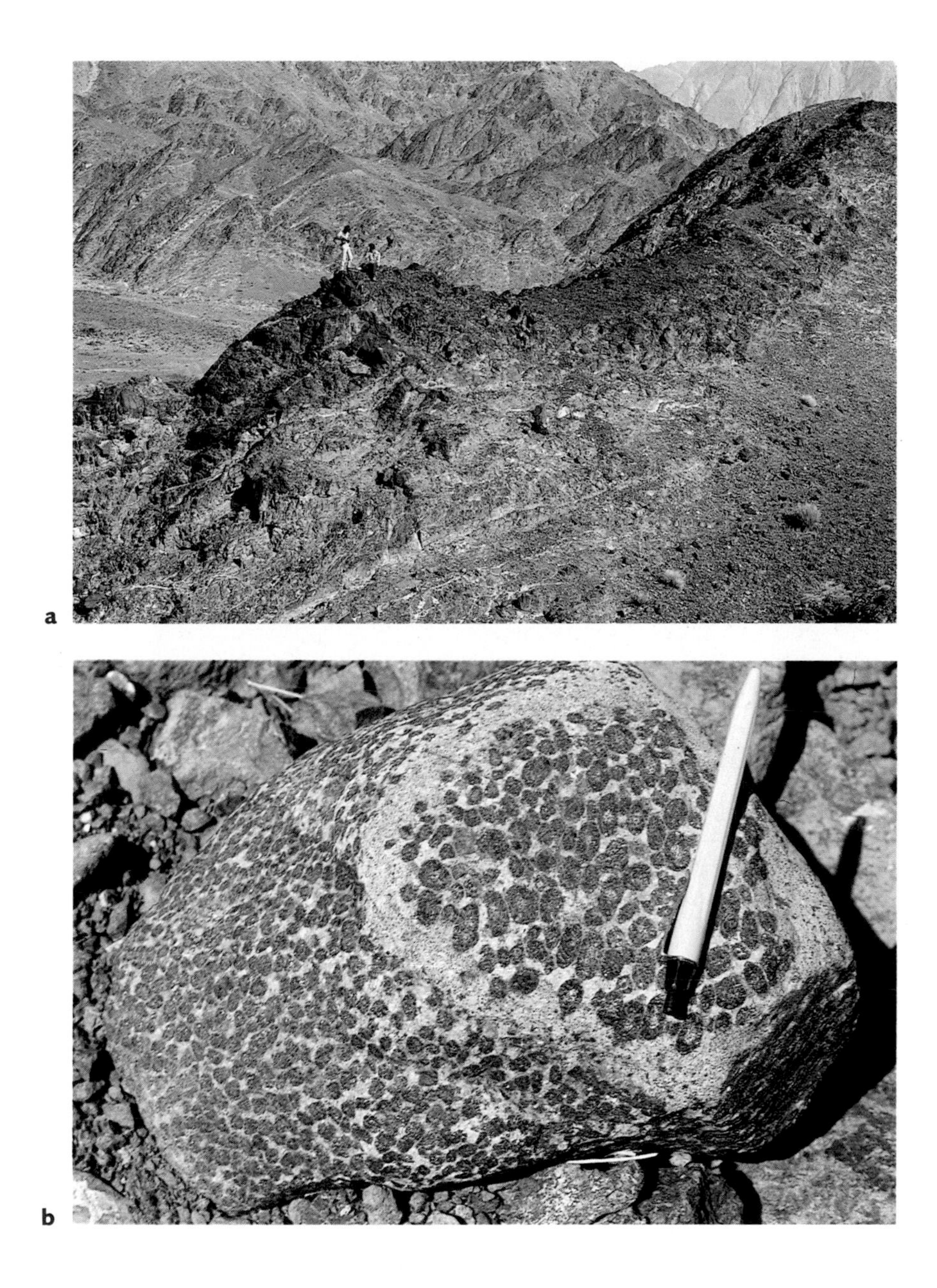

Figure 7.10 a,b
a View of a chromite deposit (*black*) in the Oman ophiolite belt. The two geologists serve as a scale for the sill. **b** Nodular minerals formed by coalescence of chromite and olivine grains in the magma. (Photos by courtesy of F. Boudier)

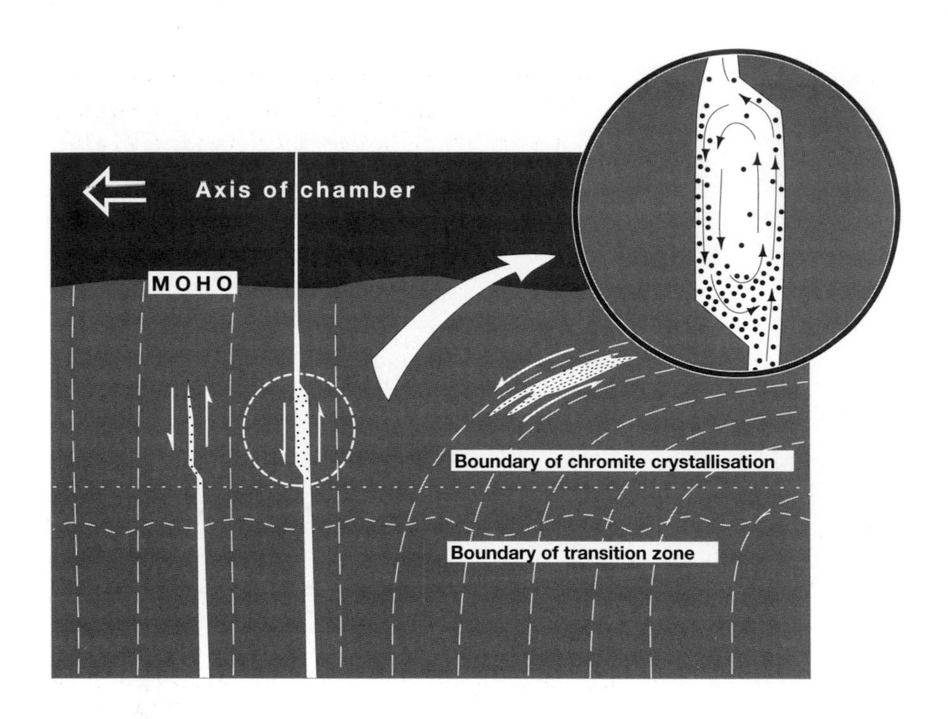

Figure 7.11
Successive stages of formation of a chromite deposit in the transition zone below an oceanic ridge. (After Lago et al. 1982, J. Petrology, 23, 103–125)

ly stirred without being able to escape from the cavity. Because of the numerous collisions, pellets of chromite (Fig. 7.10b) grow in size up to a few centimetres. They end up by becoming deposited on the floor of the cavity under their own weight. This accumulation tends to block the conduit, thereby favouring the deposition of the smaller crystals of chromite and olivine still suspended in the liquid. We have seen that basaltic intrusions through dykes operate only over a few weeks. The drop in liquid pressure at the entrance to the cavity at the end of this period also explains the deposition of the suspended crystals. A chromitite dyke will thus form within a few weeks in the transition zone of ophiolites. After filling the frequently rather irregular outline of its cavity, the chromitite dyke may become subjected to the intense deformation affecting the transition zone on its demise from the diapir. It will then become transformed into a lens with its walls parallel to the foliation of the enclosing peridotites. The complete scenario of the formation of a chromitite dyke is presented in Fig. 7.11. Other mineral deposits, like the copper sulphides formed around the black smokers on ridges, originate through their own specific mechanisms which bear witness to the wealth of natural processes.

Flattening of a Diapir Below a Ridge

Immediately below the Moho and above the diapir, the thickness of the transition zone may exceed 500 m (Fig. 5.8). From studying this particular zone we may understand how the mantle diapir will diverge laterally and what the effects of this process might be. The following scenario has been shown schematically in Fig. 7.8:

- Ascent of peridotites in the solid state, but locally also melted, taking place within a diapir, the basaltic liquid of which is progressively extracted. When the diapir reaches the level of the Moho under a fast-spreading ridge, the degree of liquid extracted will be such that the peridotite becomes refractory, forming a harzburgite.
- Retention of basaltic liquid in the transition zone immediately below the Moho. The transition zone is filled with liquid like a sponge and even to a higher degree as the peridotites may be dispersed in a basaltic brew.
- Abrupt divergence of ascending asthenosphere which becomes horizontally expelled around the margins of the diapir. The rotation from vertical to horizontal flow, both operating in the solid

state, takes place in the crystal mush of the transition zone. Calculations show that this expulsion under the pressure exerted by the diapir will lead to large velocities in the peridotites pushed away horizontally. They are displaced under the ridge at a rate which might be five to ten times higher than the drift rate of the overlying freshly created lithosphere.

- Expulsion of the liquid contained in the transition zone. As the formations of this zone move away from the axis and are taken up by the flow pushing the mantle away from the diapir, they are pressed out like a sponge to expel the basaltic liquid and fragments of peridotite which are then intruded into the overlying crust to form wehrlite bodies (Fig. 5.8). The dunites of the transition zone constitute the residues of this expulsion of magma.
- Entrainment of magma crystallizing within the magma chamber by the flow of asthenosphere diverging laterally from the diapir.
- Eventually, the exceptional, but economically important, formation of chromitite deposits along conduits through which the basaltic liquid travels from its source towards the overlying crust.

Magma Chambers: Layered Igneous Complexes?

Layered igneous complexes result from the cooling of gigantic pockets of basaltic magma which invaded the continental crust at different times. For reasons yet unknown, this took place mostly at a time when the Earth had completed about half of its present age, i. e. about 2000–2500 Ma ago. These pockets or chambers of magma extend laterally over several tens of kilometres and vertically for a few kilometres. They have been studied in detail, as among the products crystallized from these magmas one finds horizons enriched in nickel, chromium and noble metals like platinum. The mainworld producers of such minerals are actually located in those complexes. Scientists have always been baffled by the extremely regularly layered nature of the gabbros filling these chambers. These gabbros result from slow crystallization of the basaltic liquid which was injected with a temperature of about 1200 °C and then reported as solidifying at around 1100 °C. The layering is usually parallel to the floor and walls of the chamber. Close to the floor, we find the above-mentioned minerals concentrated in layers rich enough in olivine that they may form beds of peridotite. Because of this association, the concept was born that the first minerals crystallizing from the magma chamber, especially olivine and chromite, sedimented out because of their density being notably higher than that of the basaltic liquid. This explains the concentration of these early minerals in the lower levels of the chamber. The layering of the gabbros may result from this magmatic sedimentation in combination with other processes like the periodic injection of fresh basalt liquid into the chamber during ongoing crystallization or also the equally periodic development of instabilities in this liquid. Such an instability could be caused, for instance, by minerals accumulated along the cool side walls slumping into the chamber. This interpretation has been questioned and the last decade has seen a host of new theories. However, what has been retained is the concept of an accumulation of crystals formed through cooling of the liquid along the walls and over the floor of the chamber. Because of this process, the gabbros and other, similar rock facies are called cumulates.

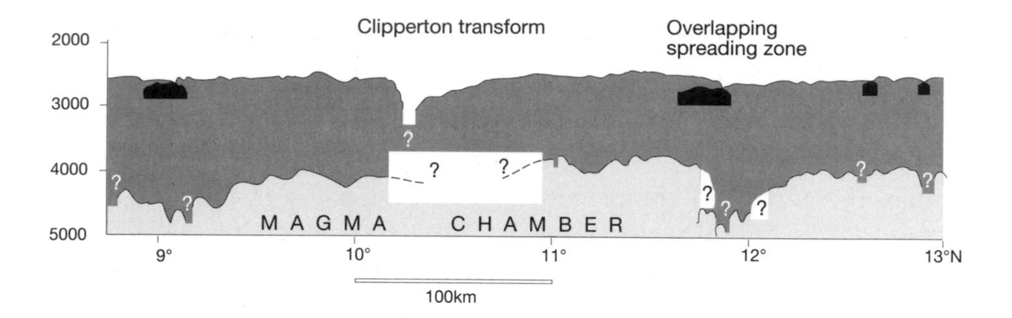

Figure 7.12
Along the East Pacific Rise, geophysicists have delineated the roofs of magma chambers, shown here in *light grey* below the darker volcanic crust. The chambers become deeper and may even disappear below overlapping spreading centres (in *black*) and fracture zones. (After R. S. Detrick et al. 1987, Nature 326, 35–41) Copyright © 1987 Macmillan Magazines Ltd.

The Magma Chambers Below Ridges

The discovery in most ophiolite complexes of layered gabbros also enriched in olivine towards the base has led quite naturally to the acceptance of the "layered complex" model. Below the ridges, the existence of vast, up to 20-km-wide chambers, was envisaged to accommodate the space necessary for this sedimentation. During the 1970s, geophysicists unsuccessfully tried to locate these chambers under active ridges and because of this failure a difficult period ensued for the proponents of the ophiolite model of ridges. Hope was restored when a few years later progress in seismic technology facilitated the recording of faint seismic echos some 2 km below the axis of the East Pacific Rise, which were quickly interpreted as indications of the roofs of magma chambers. The dimensions of the potential chambers were small, as the width of the roofs rarely exceed 3 km, and the characteristic seismic echos of the chambers were mostly completely absent along the Atlantic Ridge and far from continuous along the East Pacific Rise. With the help of further technological advances, magma chambers have now been revealed to be continuous along the East Pacific Rise over distances ranging from 50–100 km. This is apparent in Fig. 7.12, which also suggests that the interruption of the chambers coincides with the extremities of the ridge segments (Chap. 3). In the transverse section, the roof nevertheless remains narrow, but there is nothing to exclude the possibility that towards the base of the chamber the walls start to diverge, the chamber assuming a tent shape in section (Fig. 7.8).

It is not surprising that seismic data on ridges have cast doubt on the magma chamber model inspired by the layered complexes, as the similarities of their gabbro facies with those of ophiolites mask the considerable differences that exist between the two situations. The magma chamber of a layered complex possesses a cold immmobile floor whereas in the case of a magma chamber under a ridge, the floor, made up of mantle peridotites diverging from a diapir, is mobile and moreover hotter than the magma in the chamber. Any direct accumulation of magmatic sediments would be remelted and dispersed. We have furthermore shown that the peridotites below the chamber are themselves more or less liquefied, thereby constituting a highly instable floor.

Under a ridge, the roof and the side walls of a magma chamber are strongly cooled by hydrothermal circulation (Chap. 4). Consequently, the magma will start to crystallize along these contacts rather than upon the floor. We therefore would like to propose in Fig. 7.8 a model which is supported by geological data from Oman and is compatible with most oceanographic information on ridges. Seismic imagery is used to define the width of the roof. Together with numerical thermal modelling, it also confirms the field observations that the lateral walls converge upward towards the axis. Last, the width of the tectonically active zone on either side of the ridge (Fig. 3.4) allows us to estimate a width of 10–20 km for the base of the magma chamber. This is in agreement with the concept that beyond this distance the lithosphere thickens rapidly, becoming increasingly less deformable in the process. This figure of 10–20 km actually corresponds roughly to the diameter of the mantle diapirs mapped in Oman.

However, marine geophysicists identify now as a molten domain only a small lens, a few tens of metres thick, which is visible as the seismic reflector of Fig. 7.12. In cross section transverse to the ridge, it is 1–4 km wide. This would be the roof of our magma chamber. The triangular domain located below and corresponding to the bulk of our magma chamber is seismically imaged as largely crystallized, and geophysicists regard it as a solid medium. We discuss below why we think that, although largely crystallized, this is a magma, and consequently how the corresponding domain can be part of the magma chamber.

Detecting a perched melt lens is not the only result of seismic imagery at fast-spreading ridges. Attenuation of seismic waves, which is an alteration of their quality when they cross a medium imperfectly solid and elastic, suggests that the domain corresponding to our magma chamber contains only melt pockets disseminated in an essentially solid medium. At this stage, the argument can be pinned down to what is meant by magma chamber and how much melt with respect to crystals it should contain. We may define a magma chamber as a domain where flow occurs by transport of crystals within a fluid (suspension flow), in contrast to transport by deformation of crystals (solid-state flow), outside. Field data in Oman point to suspension flow taking place in the crystallizing gabbros of the large triangular domain below the ridge. In these gabbros, crystals form large laths which are remarkable piled one on the other, thus defining a strong magmatic foliation. In this well-packed medium, the fraction of liquid/solid is much reduced compared to a medium where the crystals are disoriented, just as sugar cubes are better packed in their original box than when dropped into a pot. It is possible that the layered and foliated Oman gabbros were able to flow on thin films of basaltic liquids representing the very limited fraction compatible with the seismic attenuation data.

In this model, the magma crystallizes at the base of the melt lens perched on top of the chamber, slides to the bottom to become carried along to outside the magma chamber by mantle flow which, as we remember, is expelled from the diapir at elevated rates (Fig. 7.7). This magma crystallizes increasingly against the inclined walls of the magma chamber. During this transport, the thick magma becomes dragged by the sliding movements of the underlying mantle, and the gabbros acquire their layered nature by becoming squeezed between a floor of mantle material and a roof made up of already crystallized gabbros. Despite their apparent similarities, the layering of gabbros in layered complexes and in ophiolites is thus caused by different processes. In the former, it is induced by periodic crystallization mechanisms which are still poorly understood, whereas in the latter, after the start of crystallization which may be subjected to the same phenomena, the rock becomes involved in intense deformation in the magmatic state which flattens and draws out any earlier structures into parallel lenses.

The Carapace of Ridges

From drilling close to ridges and from observations in ophiolites, we know that under the pillowed and massive lava flows on the ocean floor the sheeted dyke complex, made up of basaltic dykes which intruded into and alongside each other, extends downward towards the roof of the magma chamber. This basaltic carapace is around 2000 m thick and, at its base, the basaltic dykes extend for 100 to 200 m into the magma chamber over a particular zone, that we shall come back to. We may thus say that over the ridge itself the thickness of the lithosphere barely exceeds 2000 m. This implies that this embryonal lithosphere will not be able to resist the traction exerted by the two plates drifting away on either side of the ridge and that it will consequently rupture easily.

Over those short periods during which the crust is supplied with basaltic liquid, the crust will open up each time for about 1 m, assisted by the pressure of the magma, thereby allowing the magma to reach the ocean floor (Fig. 7.8). Over a few weeks, the magma will solidify in the fracture, a new dyke has been added to the dyke complex, and a metre of new crust created.

During the long resting periods intervening between the magmatic episodes, innumerable fissures will form. They will join up locally to form more important faults in which movements can become focussed, as shown by weak seismic activity. These fissures and open faults constitute channel-ways through which sea water can seep into the basaltic substrate. This forms the starting point of the hydrothermal circulation described in Chapter 4. Water easily penetrates towards the base of the dyke complex as the fissures follow the planes of weakness represented by the solidified sides of the dyke. After heating to 400–450 °C, this water returns to the top through strong vents, the black smokers aligned along the ridge (Figs. 4.4 and 4.5).

Between the root zone of the dyke complex and the roof of the magma chamber, the temperature increases from 400–450 °C to more than 900 °C over a distance of 100–200 m. The basaltic dykes become increasingly less distinguishable from the gabbros as the higher temperature facilitates the formation of increasingly large crystals. Furthermore, vaporized sea water locally can penetrate this zone, contributing to its cooling and initiating major recrystallization. From about 750 °C onwards, and thus to the base of this zone, the water content may initiate melting in the gabbros, leading to a first liquid which approaches granite in composition. As a result of the notoriously low potassium content of oceanic basalts, these unusual granites contain neither micas nor potassium, and are referred to as plagiogranites. From these local zones of melting, these rocks intrude into the gabbros and the overlying complex as dykes or small irregular stocks. An almost identical plagiogranite may form from basaltic magma at the end of crystallization in a magma chamber after the minerals crystallizing at high temperatures have been removed.

The Reconstructed Ridge

How could we reconcile the narrow and probably turbulent ascent of the mantle flow represented by the diapirs with the expansion of fast-spreading ridges, which takes place in a regular fashion, continuous in space and time? Could there be any relationship between the spacing of the diapirs in the mantle and the longitudinal segmentation of the ridges which represents the only clear sign of some degree of discontinuity in the operation of a ridge?

Although the ridge activity is basically controlled by the turbulent ascent of mantle, closer to surface the lithosphere may throttle this activity by its rigid nature. Let us recall that lithosphere forms along the ridge by cooling of crust and mantle and that it thickens in a regular fashion as it moves away from the axis at an equally regular rate (Fig. 1.7). We have also examined the role of fracture zones, the vertical walls of which contribute to channeling the movement of asthenosphere expelled from the diapirs.

Let us start from the centres supplying basalt to the crust, viz the mantle diapirs. In Fig. 7.8, which compiles our knowledge and hypotheses on the subject, we note that in the area just above the diapir, the magma chamber may be somewhat pinched by the pressure from the diapir. The basaltic magma expelled laterally will precipitate layered gabbros at some distance from the diapir. The magma thus flows along the axis, indirectly feeding the chamber over a distance of several tens of kilometres on either side of the diapir. A sudden and violent intrusion of basaltic liquid breaks open the roof of the magma chamber, forming a new dyke in the dyke complex and feeding a lava flow on surface. Within the dyke, magma mostly advances horizontally along the ridge. The horizontal propagation of these dykes along the axis may be followed on the emerged ridge on Iceland by small seismic tremors marking the advance of the fracture. After having covered a distance of 10 km in 24 h, the dyke may attain a distance of 30 km from the point of injection within only a few days.

Based on these data, Fig. 7.8 shows how a diapir which itself extends over only about 15 km along the ridge could nevertheless supply the crust with basaltic material over several tens of kilometres because of the flow of magma along the ridge within the magma chamber as well as within the dykes of the dyke complex.

Thus each diapir under a fast ridge could be the source of material for a full ridge segment whose length (50–100 km) is defined by its overlapping spreading centres (Figs. 7.5 and 7.12). The ends of the adjacent segments, which one could interpret as "competing" with each other, are rather mobile, an expression of the variable activity of the magma chambers. When the chamber is strong, it will advance by pushing ahead its extremity, which passes, wraps around and sometimes even penetrates the end of the adjoining, less active segment. The two segments may thus communicate, at least on a temporary basis. We may reconstruct past duels between two such competitors from the scars which they left on the oceanic crust formed over the last 2 million years.

The topography along fast-spreading ridges reflects the internal activity, as the highest points coincide with centres of segments whereas the lower-lying parts mark their end portions (Fig. 7.12). The central uplift would occur right above the diapir. The pressure from the diapir and the dilatation resulting from the heat flow and the ascent of basalt contribute to raising the relief against the extremities of the segments where these effects are absent or only feebly developed.

8 From Rifts to Fast-Spreading Ridges

The expansion of continental rifts is of the fissural type: a fracturing in the lithosphere causes a wedge of hot partially molten asthenosphere to ascend from the mantle. Basaltic liquid drawn from this wedge fills the fissures in the crust, creating a swarm of dykes. This episodic activity progressively heats the lithosphere, making it weaker and allowing it to respond by tectonic stretching, the second type of expansion. In this way we proceed to oceanic rifts like the Red Sea framed between two stretched continental margins. When the so far rather low spreading rate increases to more than 2 cm/a, the continuous creation of oceanic lithosphere will lead to the relatively regular structure of fast-spreading ridges (Chap. 7) characterized by the harzburgitic ophiolites. Let us sum up the concepts inherent in this expansion: rapid rise of asthenosphere, little loss of heat, strong melting, highly depleted nature of the residual mantle (harzburgitic), copious and constant supply of basaltic liquid to the crust, development of magma chambers, crustal thickness 6 km. When, in contrast, the rate of oceanic expansion is low, formation of crust may become discontinuous, as suggested by recent studies along the Atlantic Ridge and ophiolites of the lherzolite-type form. There is a succession of episodes of "normal" basaltic supply, possibly caused by the uprise of a mantle diapir with intervening periods, during which tectonic stretching may lay bare the mantle on the ridge. On a global scale, the slow or discontinuous ascent of asthenosphere diapirs leads to a lower degree of melting (mantle residue lherzolitic) and thus to a lower rate of basalt formations (thinner crust, episodic presence of magma chambers). The spreading rate thus appears to be the key factor controlling the type of mantle accretion below the ridges as well as their segmentation. Differences in the temperature of the rising asthenosphere should also be considered.

Slow- or Fast-Spreading Ridges and Ophiolites

The study of ocean floor topography as outlined in Chapter 3 has told us that there are important differences between fast-spreading ridges on the one hand and slow-spreading ridges and rifts on the other. The "comparative ophiolitology" of the end of Chapter 5 highlighted the distinction between a "harzburgitic" and a "lherzolitic" type. We shall now stress the correspondence between ridges and ophiolites and show that in a general way the harzburgitic ophiolites are derived from fast-spreading ridges and the lherzolitic ones from slow ridges. This correlation permits a better understanding of the functioning of the different types of ridges and an overview of the system of oceanic expansion.

In the preceding chapter, our attention was focussed on the functioning of fast-spreading ridges. After now examining the functioning of slow-spreading ridges and rifts, we shall study how and why the rate of expansion appears to be the main cause for the differences in the system of oceanic expansion. Other factors, which include the proximity of mantle plumes such as Iceland, the geodynamic environment or the age of the expansion system, appear to play a local or minor role. Actually, the information furnished by oceanographic studies on the expansion of the mid-oceanic type and on that of the back-arc basins does not reveal systematic differences tied to the geodynamic environment. As far as age is concerned, the comparative study of ophiolites of greatly differing age, the oldest being close to 2000 Ma and the youngest virtually contemporaneous does not show any notable differences. This is by no means surprising, as we know that the global tectonic system has been operating according to the same rules over at least the same time span. This, however, does not exclude the possibility of cyclic activity in the Earth system, a question that ophiolites may help to answer (Chap. 9).

Expansion of Rifts

The expansion of an oceanic rift like the Red Sea is comparable to that of a slow ridge creating oceanic lithosphere, with a spreading rate in the order of 0.6 cm/a. Before looking into this situation, let us briefly examine continental rifts, the expansion of which is still slower, with rates in the order of 0.5 cm/a or even less.

For continental rifts we do not possess sections comparable to those offered by ophiolites and, as outlined in Chapter 3, our knowl-

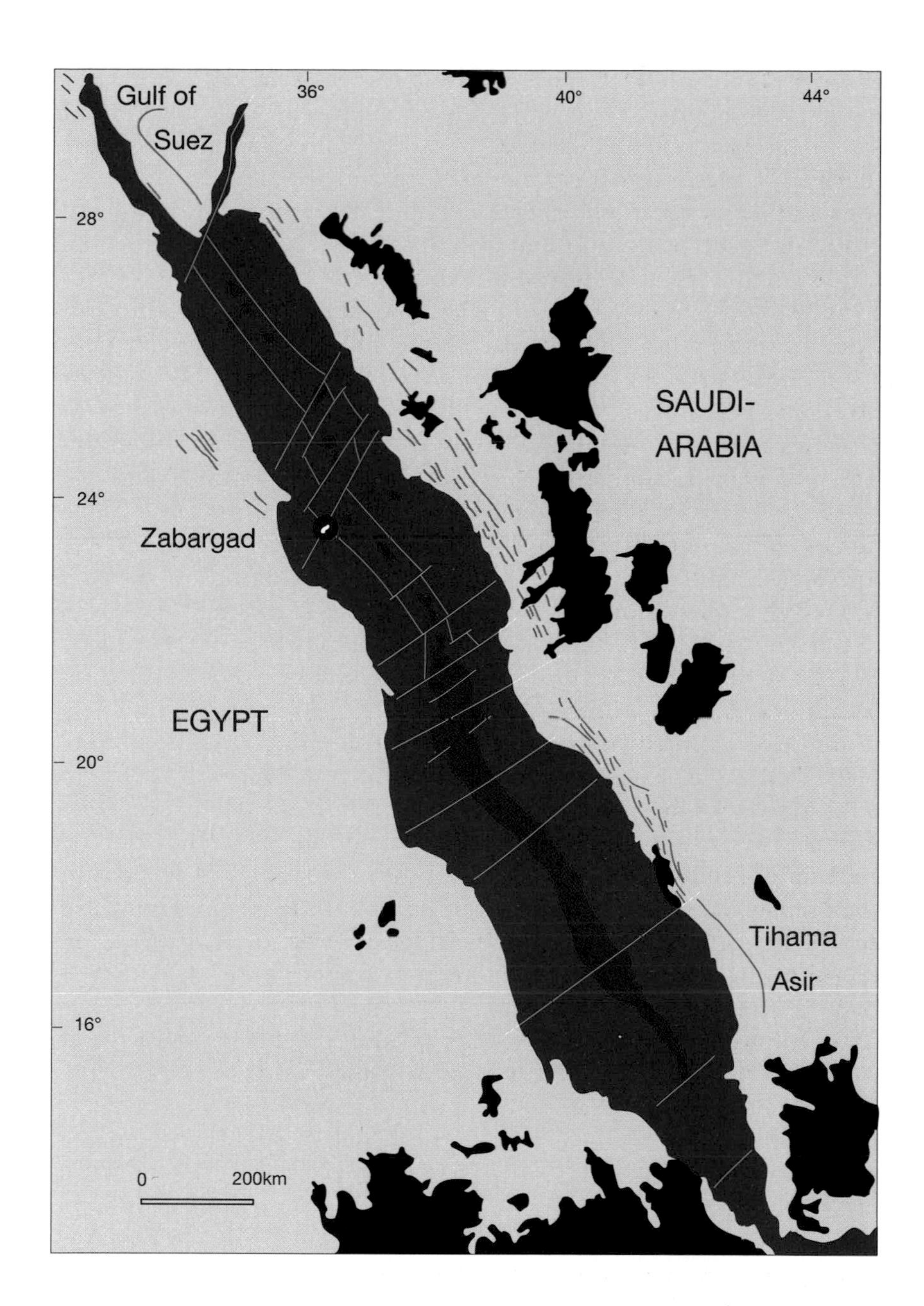

Figure 8.1
The Red Sea, a rift on the path to "oceanization". The continental crust which was thinned 5–10 Ma ago (*light red*), borders on oceanic crust (*dark red*) created along the axis. Fissure volcanism (flows in *dark blue*, dykes in *green*) and the intrusion of a mantle wedge on the island of Zabargad some 20–25 Ma ago prepared the present situation

edge is based solely on surface structures and major geophysical trends. The geology of passive margins is known better because of the oil and gas potential of these areas. We here recognize the role played by faults in stretching the first kilometres of continental crust. A passive margin represents the final stage of continental rifting and is thus not necessarily representative of an intracontinental rift which would correspond to a less evolved situation.

When flying above arid countries, a traveller frequently notices from his window innumerable parallel basalt dykes. Similar dykes are also evident on geological maps like those from northern Brittany or Scotland. The intrusion of these dykes through continental crust bears witness of rifting activity. Thus the Arabian coast of the Red Sea is pervaded by basalt dykes aligned parallel to the coastline (Figs. 8.1 and 8.2). These were injected into this continental crust some 20–25 Ma ago during an episode of continental rifting, preparing the way for the formation of oceanic crust of the "oceanization" of the Red Sea which started about 5–10 Ma ago or about 15 Ma after the continental rifting phase. In this case, the expansion of the rift was assured by the injection of dykes and its absolute amount may be estimated from the cumulative width of these dykes. Locally, as at Tihama Asir, these basalt dykes constitute within the continental crust a dyke complex overlying a sequence of layered gabbros produced by the crystallization of basalt in a magma chamber. This "ophiolitic" section is evidence for pronounced local expansion. On the other hand, in the continental crust of Egypt, which is thinned out along the Red Sea, the same expansion 20–25 Ma ago appears to have resulted in a local rupturing of the crust. This is suggested by investigations on Zabargad Island, where a mantle wedge cuts across and deforms the continental formations (Fig. 8.3).

In continental rifting thus two expansion mechanisms may be at work: fracturing with injection of virgin material derived from the mantle (Fig. 8.4A) and stretching along inclined faults on surface, grading into plastic deformation at depth (Fig. 8.4b). The fracturing envisaged here operates on a crustal scale through intrusion of basalt dykes, but also on a lithospheric scale by way of intrusion of mantle wedges into the ruptured lithosphere. In this concept we envisage that rifting of normal, i. e. cold and resistant, lithosphere as a first

Figure 8.3
Lithospheric "fracturing" with an ascending asthenospheric mantle wedge and injection of basaltic dykes in the crust. Model inspired by studies of Zabargad Island. (After A. Nicolas et al. 1987, J. Geophys. Res. , 92, 461–474) ▶

Figure 8.2
Swarm of basaltic dykes (*black*) along the Arabian shore of the Red Sea in North Yemen. Their common trend is parallel to that of the Red Sea. (cf. Fig. 8.1; photo by courtesy of G. Féraud)

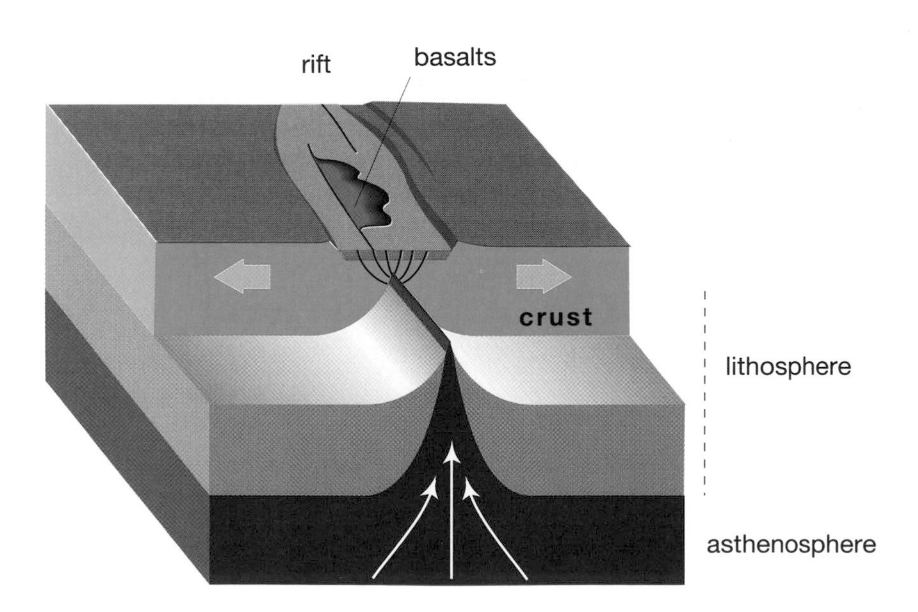

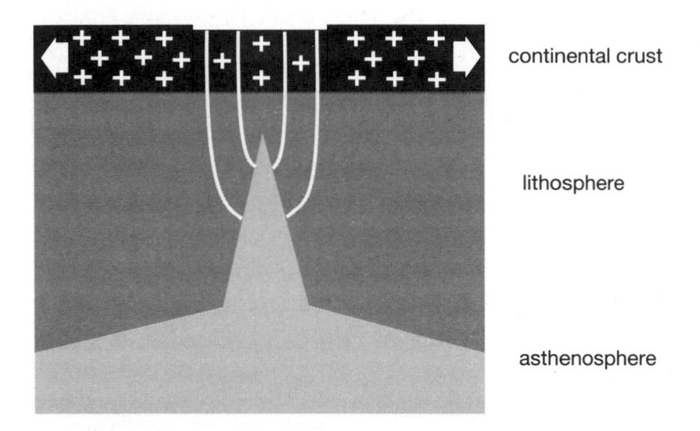

A - Expansion by injection

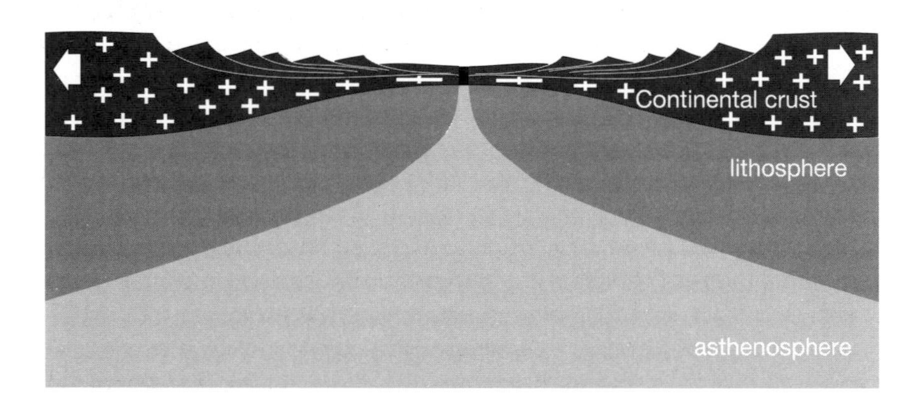

B - Expansion by tectonic stretching

Figure 8.4 A,B
The two modes of lithospheric expansion. **A** Injection of hot melting mantle (asthenospheric) into the cool and fractured mantle of the lithosphere, and injection of basalt dykes into the crust. **B** Tectonic extension with faults (listric faults) on surface and plastic deformation at depth. (After A. Nicolas et. al., 1994, E. P. S. L., 123, 281–298)

step will lead to expansion by fracturing and intrusion of mantle-derived material. This took place in the Red Sea, which was still continental 20–25 Ma ago. In a second step, after a time lag of 15 Ma, the lithosphere was sufficiently reheated to become plastic and started to become stretched. Thin passive margins started to develop and oceanization became possible.

Expansion of Slow-Spreading Rifts

The Moho Uncovered

A very important discovery in recent years in the oceanic crust of slow ridges, repeated at several sites along the Mid-Atlantic Ridge and in the Tyrrhenian Sea, concerned the ovservation of direct outcrops of mantle peridotites or their presence immediately below sediments and basalts. This has been a unexpected discovery as the detailed observations carried out by French and American geologists during the FAMOUS campaign of 1974 had shown that the axial valley of the ridge and its flanks were uniformly covered by basalt. The fact that dredgings over transform faults regularly furnish peridotite samples had been explained by these faults cutting through the entire crust and penetrating down into the mantle. Instead of moving only horizontally along the fault (Fig. 3.6), one of its flanks could have been subjected to a certain upward component thereby exposing the mantle peridotites, especially as the crust would be particularly thin in the vicinity of these faults. However, how could we explain the, admittedly local, presence along the same ridge of mantle peridotites on the sea floor instead of below 5–6 km of newly formed basaltic crust? In these areas, boreholes and dredgings have also recovered many samples of extremely deformed gabbros, and observations from submersibles revealed that faults were actively shaping the relief of slow ridges as in rift situations (Fig. 3.14). There is thus a very important tectonic activity along the slow-spreading ridges.

The Rhythms of Slow-Spreading Ridges

The contrast between sites along the ridge on which crust is completely or partly missing and others, like the FAMOUS sites, on which the crust appears to be normal, suggests a discontinuous activity of these ridges. By definition, the ridges are sites along which new lithosphere is created by oceanic expansion. The study of rifts told us that expansion may be accomplished either by tectonic stretching as in the case of passive margins by fault activity and plastic stretching or by the intrusion of basaltic magma into parallel fractures in the rift (Fig. 8.4). We shall see that the expansion of slow ridges takes place through the alternation of both processes. Whereas in fast-spreading ridges the abundant and regular supply of basalt continually creates new crust occupying the space opened by expansion, it appears in slow ridges that the supply of basaltic magma is rhythmic, with a periodicity of 1–2 Ma (Fig. 8.5). When supply is normal, the ridge operates like a fast-spreading ridge developing in particular a magma chamber under the usual shield of lavas and the dyke complex. When the supply of basaltic magma fades away, tectonic extension starts to replace the supply of magma in order to maintain the spreading of the system. The chamber solidifies and the still hot gabbros resulting from the ensuing crystallization suffer a considerable degree of stretching and are transformed into the notorious highly deformed facies observed in the corresponding slow spreading oceans. The new crust is then made up only of lavas and the dyke complex, and its thickness will amount to only half the normal value. When the magma supply stops entirely, this cover of lavas and dykes may split open along large and highly active faults which uncover the underlying mantle, allowing it to come to the surface now along the axis of the ridge.

The study of ophiolites of lherzolite type confirmed in detail the operation of slow ridges. We have seen especially in the case of the Trinity ophiolite (Chapt. 5) that the magma chambers were not continuous and that intense deformation may affect the gabbros penetrating even down into the peridotites. Other ophiolites of this lherzolite type, like the Tibetan ophiolite of Xigaze, are practically devoid of gabbros, the basalt dykes rooting directly into the peridotites. Finally, the ophiolites of the western Alps furnish us with the example of an ocean floor stripped of its crust (Fig. 8.6). This floor belongs to the ocean which separated the European plate during the Mesozoic from a promontory of the African plate destined to collide with it. Reconstructions are rather intricate here because of the intense tectonism of the Alps, but they nevertheless show that along faulted contacts, the lherzolitic mantle in certain places cropped out directly on the ocean floor, being locally intersected by basalt dykes or covered by basalt flows and slivers of highly deformed gabbros (Fig. 8.6).

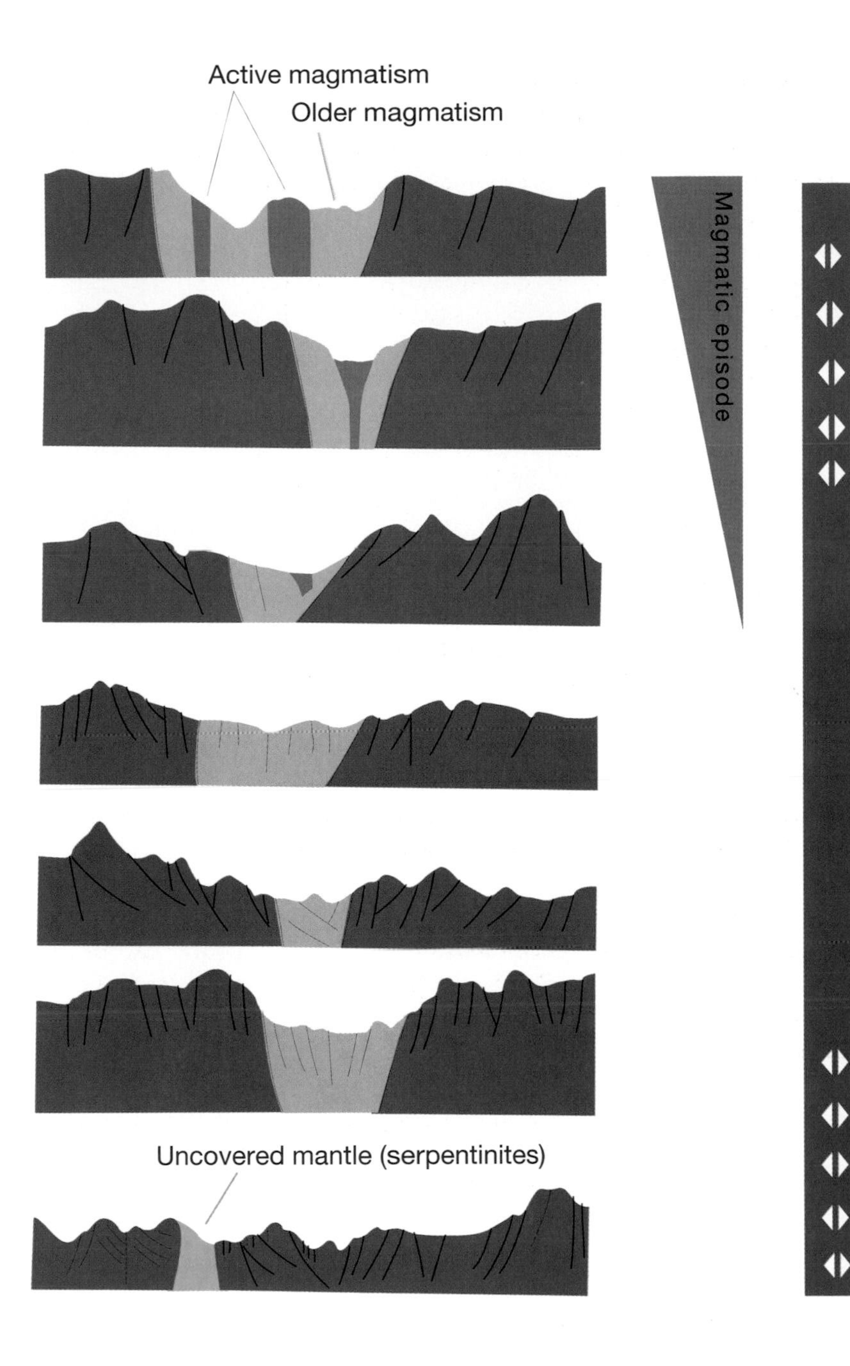

Figure 8.5
Pulses of magmatism and tectonic extension on slow ridges, illustrated by a series of sections across the Mid-Atlantic Ridge. (After J. A. Karson et al. 1987, Nature, 328, 681–685)

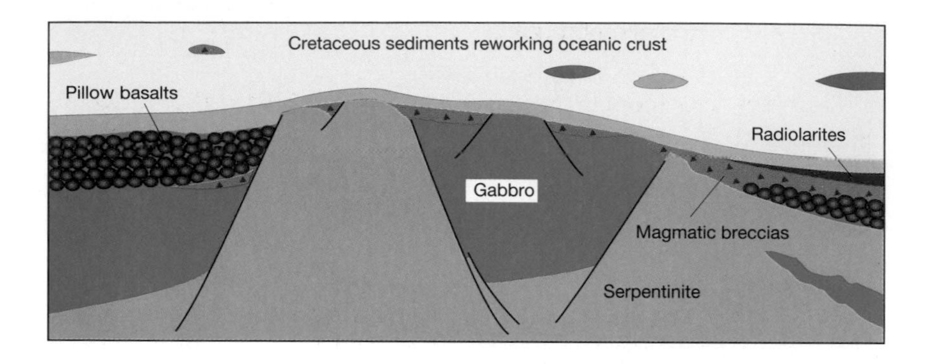

Figure 8.6
Reconstruction of the floor of the ocean separating Europe from Italy prior to the alpine collision, based on studies of the ophiolites scattered throughout the mountain range. The mantle furnishing these ophiolites belong to the lherzolite type, suggestive of a slow ridge environment. Locally, oceanic crust must have been missing or was reduced to a few hundred metres of basalt only. (After P. Tricart and M. Lemoine 1986, Tectonics, 5, 95–118)

Is the Moho Really at 6 km?

How can we reconcile the presence of the Moho at 5–6 km, as known from seismic data, under the crust below most of the oceans with the existence of variable crust thickness or its absence under slow ridges? We are certainly aware that the depth of the Moho is not uniform at 5–6 km. Its depth is only 4 km close to transform faults and along slow-spreading ridges, spreading at a rate of below 2 cm/a like the ridge in the Arctic Ocean. This does not solve the problem. Let us recall that the Moho is a seismic entity. Along this level the propagation velocity of seismic waves increases abruptly, making it the excellent reflector it represents. The base of the "normal" oceanic crust, along which the gabbros are in contact with the peridotites, also possesses the desired contrast in velocity and it would be legitimate to correlate this layer with the geophysical Moho. However, strong serpentinization of mantle peridotites will also lead to a pronounced drop in seismic wave velocities, from 8.2 km/s to 7 km/s and less. This serpentinization is caused by the penetration of sea water into the peridotites along fractures controlling hydrothermal circulation. The limit of serpentinization is thus determined by the depth below which the rock pressure would no longer permit open fractures. We know from experiments that this depth does not exceed 6 km although there will be some variation depending on temperature and ambient constraints. It is thus conceivable that in some areas, and especially in the outcrops of serpentinites on ridges, the Moho is located at the serpentinite-peridotite boundary. In lherzolitic ophiolite massifs in which the thickness of the crust is less than 5–6 km, we observe a high-temperature serpentinization of the peridotites which is attributed to ridge hydrothermalism. This type of serpentinization is virtually unknown in ophiolites of the harzburgite type, in which the crust is of a thickness that normally inhibits the contact of mantle with the ocean waters.

In Search of a General Model

Continental rifts with low opening rates obtaining new material only from volcanoes and their feeding dyke swarms, slow-spreading ridges with thin or even absent basaltic crust and corresponding ophiolites of the lherzolite type, and finally fast-spreading ridges with thick basaltic crust and ophiolites of the harzburgite type: they all highlight the obvious relationship between spreading rate and supply of basaltic magma from the mantle. However, from a rate of 3–4 cm/a up to nearly 20 cm/a, viz. the spreading rate of the fastest ridges, the crust will not thicken further. The study of the mantle (Chaps. 6 and 7) furnishes us with the explanation to this situation.

Let us summarize how the mantle melts under normal conditions, that is away from hotspots (Fig. 6.6): asthenosphere rising below the spreading centres will start to melt when it crosses a depth of 75 km. The fraction of basaltic melt thereby produced will increase regularly as the ascent continues. Melting will stop, however, when this asthenosphere cools in contact with the lithosphere in which it will eventually become incorporated.

It is the spreading rate which controls the depth attained by the asthenosphere charged with basaltic liquid below the centre of expansion or, using another term, the thickness of the lithospheric "barrier" stopping this ascent. The lithosphere actually becomes thicker with age just like a crust overlying a bath of molten paraffin (Fig. 1.7). When the expansion is slow, the lithosphere will be thick in the vicinity of the rift or the ridge, whereas it will be thin when expansion is fast. We conclude that below a rift, the asthenosphere stops at around 30 km in depth – yielding only a moderate fraction of basalt. Under a fast ridge, however, the asthenosphere will rise very high to form a magma chamber, and it delivers a maximum volume of basalt. We observe that the asthenosphere achieves its highest level, i. e. the crust when the spreading rate is above a few centimetres per year, a situation considered as fast and that the situation stays like this at even higher rates. This conclusion is drawn from a study of the diagram in Fig. 6.6 and is presented in a more direct form in Fig. 8.7. Let us now examine how the field data on peridotites associated with these different settings might support our analyses.

The Signature of the Mantle

With peridotite massifs as the only evidence, we can suggest that the collision responsible for stacking up the Pyrenees mountain range, at least in the eastern part of the belt, followed the emplacement of small intracontinental rifts. In contrast to this, the collision of the western Alps followed the closure of an ocean that had been spreading at a slow rate. These conclusions are derived from Fig. 8.7 and are based on the presence of numerous small spinel-bearing lherzolite massifs in the eastern Pyrenees and on plagioclase-bearing lherzolite massifs associated with ophiolite remnants in the western Alps. We have already pointed out the relationship between the harzburgitic nature of certain ophiolites and fast-spreading ridges (Fig. 8.7a) and between plagioclase-bearing mantle and slow-spreading ridges (Fig. 8.7b). Figure 8.7c implies that the peridotites underlying a continental rift are spinel-bearing lherzolites. The corresponding explanation is offered by the general model: when the ascent of asthenosphere comes to a halt at a depth of 30 km or more, as under a rift, the peridotite will undergo melting within the stability field of spinel-bearing lherzolite. It thus becomes incorporated in the surrounding mantle as a spinel-bearing lherzolite. For this reason, the vast majority of mantle nodules brought up by basalts so frequently in certain rifts volcanoes are spinel-bearing lherzolites. When the ascent is stopped at a depth of about 15 km as under slow-spreading ridges, the peridotite will equilibrate in the field of plagioclase-bearing lherzolites; and when, eventually, the ascent continues right into the crust as under fast-spreading ridges, the still more pervasive melting has consumed all the original clinopyroxene and the residual peridotite will be a harzburgite.

The mineralogical facies of peridotites thus reflect the type of spreading environment. We shall see now what the structural orientation of the same peridotites can tell us.

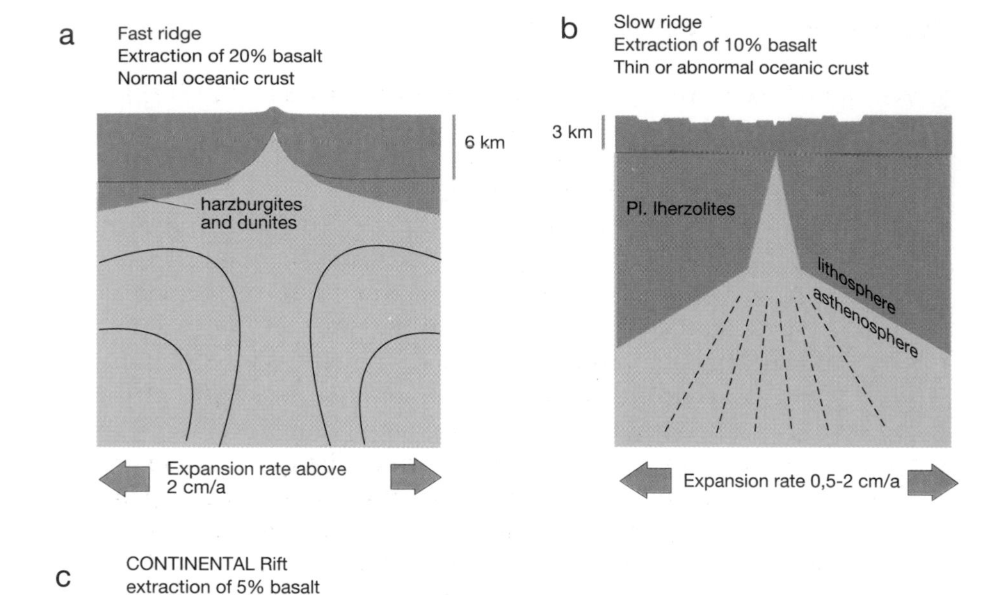

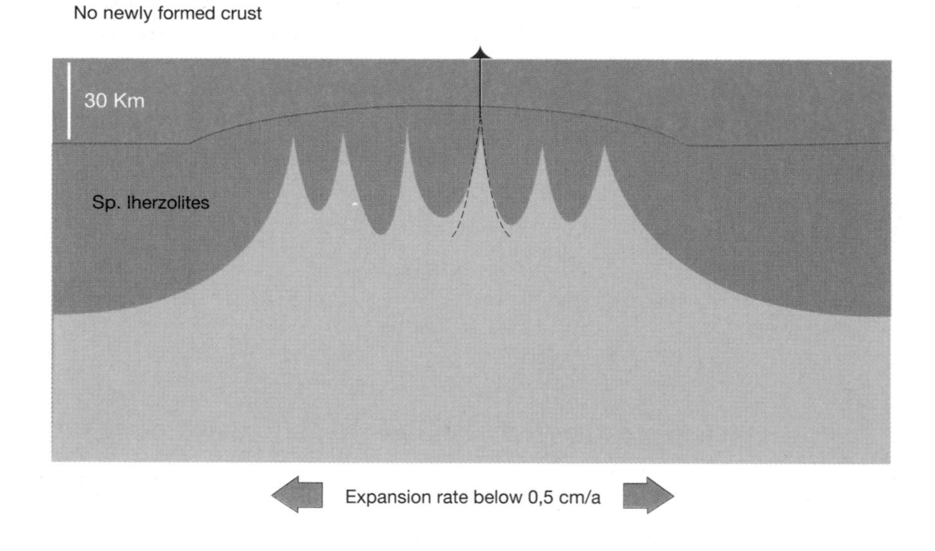

Figure 8.7 a–c
Fast-spreading ridge (**a**), slow-spreading ridge and oceanic rift (**b**) and continent rift (**c**) the key role played by expansion rate. The model of the continental rift was inspired by the Massif Central of France. (After A. Nicolas 1986, Rev. Geophys., 24, 875–895)

The Wedge and the Mushroom

We know how to recognize in peridotites the so-called asthenospheric deformations which are induced by ascent and spreading. We know also that closer to the lithosphere the shear flow of asthenosphere is guided by the rigid underside of the lithosphere (Fig. 6.5) and we have concluded that studying the trace of these shear planes (the foliation) will permit us to determine the orientation of the plane separating lithosphere from asthenosphere.

In peridotites of harzburgite-type ophiolites (Fig. 5.8), the foliations are parallel or near-parallel to the Moho, showing that the lower surface of the lithosphere is only very slightly inclined, a characteristic feature of fast ridges (Fig. 8.7a). In contrast to this, the inclined foliations in peridotites of lherzolite-type ophiolites (Fig. 5.8) show that the lower side of the lithosphere is dipping at a higher angle, this being an indication for slow spreading (Fig. 8.7b). The sharp dip contrast envisaged here between fast and slow spreading ridges is larger than that predicted by the respective dips of the thermal lithosphere (~ 1000 °C surface). This is so because we are dealing here with the mechanical lithosphere whose behaviour is highly sensitive to small temperature changes.

This analysis confirms the correlation between harzburgitic ophiolites and fast-spreading ridges and between lherzolitic ophiolites and slow-spreading ridges or oceanic rifts; but we can proceed even further.

Lithospheric Underplating

Since the start of plate tectonics, debate has been incessant about the way that asthenosphere might rise under the ridges and accrete to form new lithosphere. One model entails a thick lithosphere cut by a wedge of asthenosphere which rises under a ridge somewhat like the situation shown in Fig. 8.7b. Another model implies that the lithosphere is very much reduced in thickness below the ridge and is thickened regularly at either side (Fig. 8.7a). Field studies in peridotites show that the two models are not mutually exclusive and that they correspond to rifts and slow ridges on one side and to fast ridges on the other. Thus, accretion of lithosphere under a rift or a slow ridge will take place by intrusion of a wedge of asthenosphere which is plated laterally against the adjacent lithosphere. Under a fast ridge, asthenosphere rises like a mushroom or diapir which will diverge, and accretion takes place by underplating against the overlying lithosphere.

It must be kept in mind that the schematic drawings of Fig. 8.7 represent only the extremes and that there are many intermediate stages. In particular, the analysis does not cover ridges located over a hotspot as on Iceland. This is a special situation in which, in the environment of a slow ridge like the Atlantic one, the mantle is particularly hot and rises probably right up to the crust, undergoing pronounced melting on the way. It is, moreover, this contrast between slow spreading and the vast supply of basaltic liquid which here leads to the formation of crust with 10–20 km thickness instead of the usual 5–6 km.

Mantle Diapirism as the Cause of Segmentation and Periodic Activity of Ridges

Diapirism and Segmentation

In the preceding chapter we proposed that each segment of a fast-spreading ridge which is limited longitudinally by overlapping spreading centres is associated with a mantle diapir which feeds it by way of a magma chamber extending over several tens of kilometres along this ridge (Fig. 7.8). From this situation we may conclude that the mean spacing of diapirs in the mantle is in the range of 50–100 km, the same as the mean length of the segments along the East Pacific Rise. Adjacent diapirs revealed by field mapping in the Oman ophiolites are actually about 30–50 km apart. However, diapirs represent mantle floor instabilities; essentially, this precludes one from being too square in this kind of prediction. It may well be that certain segments are fed by one single large diapir and others by a number of smaller diapirs.The diapirs are derived from a mantle layer at a depth around 50 km which is instable because of melting. Their spac-

Figure 8.8
Gravity "bulls eyes" along the Southern Mid-Atlantic Ridge (*below*) and their relation to the segmentation defined by transform faults (*above*). These gravity anomalies, seen at Moho level by "striping off" the crust by calculation, are thought to represent the contour of low density mantle diapirs. Notice that each diapir would coincide with the center of a segment. (After Neumann and Forsyth, 1993, J. Geophys. Res. 98, 17: 891–910) ▶

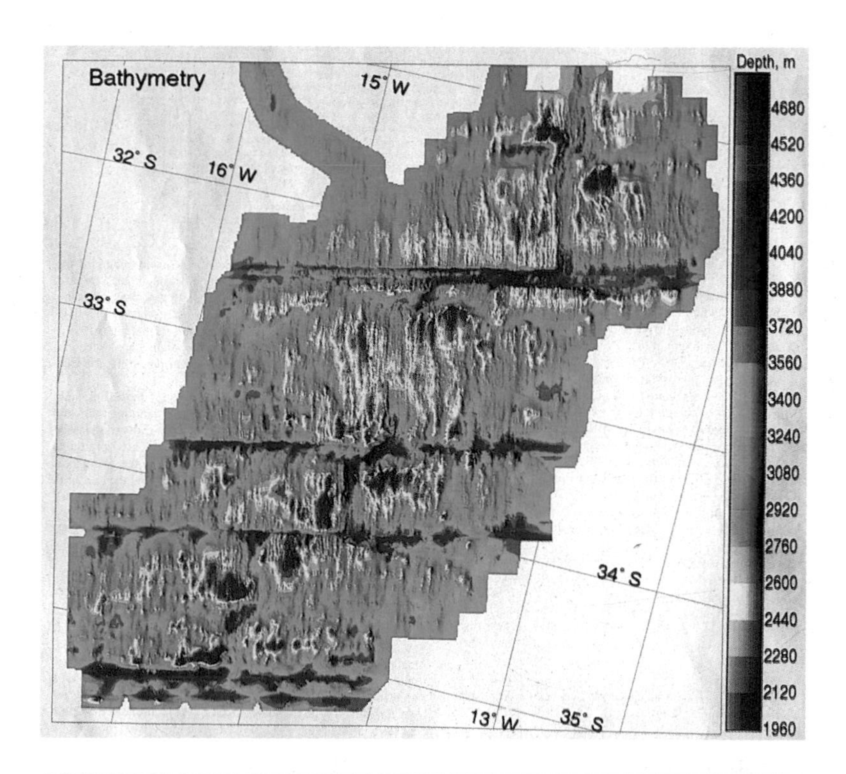
Bathymetry
Depth, m
15° W
32° S
16° W
33° S
34° S
13° W
35° S
4680
4520
4360
4200
4040
3880
3720
3560
3400
3240
3080
2920
2760
2600
2440
2280
2120
1960

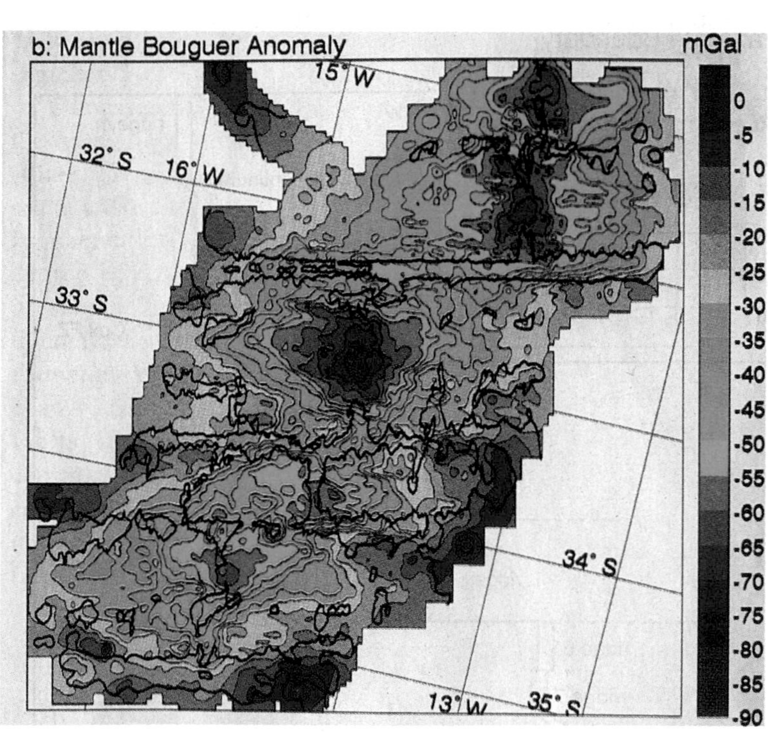
b: Mantle Bouguer Anomaly
mGal
15° W
32° S
16° W
33° S
34° S
13° W
35° S
0
-5
-10
-15
-20
-25
-30
-35
-40
-45
-50
-55
-60
-65
-70
-75
-80
-85
-90

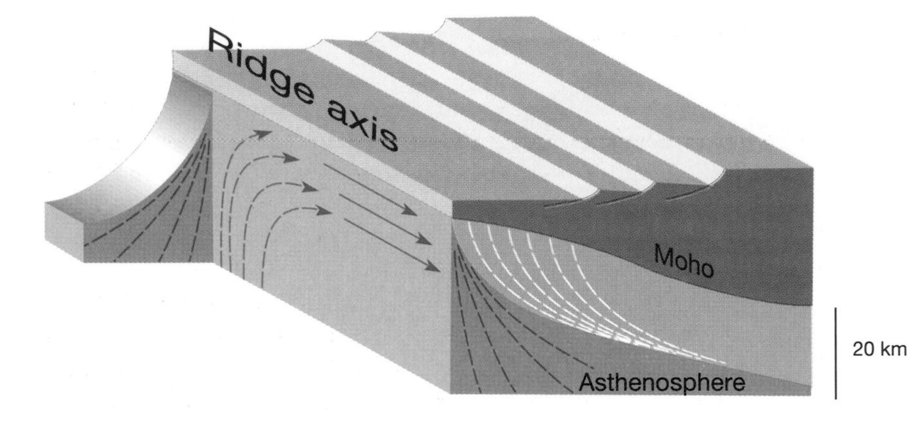

Figure 8.9
Mantle diapir rising below an oceanic rift like the Red Sea. Movement in and near the diapir is channelled by the relatively steep walls of the lithosphere below the rift. (After F. Boudier et al. 1989, Geol. Soc. Am. Bull. , 110, 820–833)

ing is controlled by the properties of this unstable layer and thus barely subjected to the thickness of the lithosphere in the vicinity of the ridge and with this to the spreading rate. We can thus expect that the spacing of diapirs should be more or less the same under fast- and slow-spreading ridges. The segmentation of the Atlantic Ridge, our model of a slow ridge, is marked by fracture zones with a mean spacing of 50–100 km. One could thus be tempted to suggest that the segment length is controlled by a comparable spacing of the diapirs rising under the ridge. This conclusion agrees well with the estimates outlined above. It has been confirmed recently by gravimetry studies along this ridge which have revealed gravity anomalies thought to represent mantle diapirs (Fig. 8.8).

Whereas the situation near the layer from which the diapirs are derived should not differ much between the various types of environment, this does not apply to the zone of interaction with the lithosphere. We have talked in the preceding paragraph about the mushroom-like rise of diapirs under fast ridges and the wedge-shaped rise under slow ridges. Field studies in lherzolitic ophiolite massifs, comparable to the slow ridge model, have shown that the asthenospheric rise within the wedge took place at angles differing from the theoretically expected vertical ascent. This suggests that the ascending flow of the diapir is channelized by the lithospheric wedge with a dominant flow component parallel to the axis of the rift or ridge (Fig. 8.9).

The Diapirs: Tubes or Blobs?

Calculations on the ascent of diapirs indicate that it would take less than 0.5 Ma for a diapir to arrive below a ridge from the starting layer deeper down. Unfortunately, these calculations do not say whether the diapirs represent conduits with a lifespan of several million years or more, with a tube-like shape through which partially molten asthenosphere rises in a more or less continuous fashion, or whether they behave like blobs detaching themselves from the unstable layer and ascending one after the other. The scars left in the several million years old lithosphere of the East Pacific Rise by the duelling activity of the extremities of the segments exhibit indications which are rather interesting in this respect.Actually, the advances and retreats of the extremities are only the expression of minor pulsations in supply whereas the segments themselves possess a comparable duration. We may thus conclude that the diapirs supplying these segments possess a similar lifespan of several million years. They would thus be "tubes".

In contrast to this, the 1–2 Ma rhythms of magmatic supply along the Atlantic Ridge are most easily explained when we assume that the diapirs assuring the supply rise like blobs.The ascent of such blobs (less than 0.5 Ma) would correspond to a magmatic episode, followed by a period of tectonic extension preceding the next magmatic episode some 1–2 Ma later.

9 The Major Pulsations of the Earth

Was the existence of Supercontinent Pangea some 200 Ma ago the random result of erratic plate movements or should it be considered as a product of deeper forces? Several factors are in favour of the concerted action of forces which, combining and dispersing the successive Pangeas, shaped the Earth in rhythms covering some 400 Ma each. The presence of such Pangean cycles could explain why no ophiolites appear to be forming at present in the oceans. They grounded on the continental margins within the "lithospheric tidal flats", i. e. during the most active phase of dispersion of a supercontinent. This phase occurred about 100 Ma ago for the most recent cycle and about 500 Ma ago in the case of an older, still more hypothetical, cycle.

A Geological Police Investigation

Let us imagine that in a perfectly peaceful world in which the concept of aggression has become completely eradicated, a unique incongruous crime has been committed. As crime no longer existed, the policemen entrusted with the investigation would be rather embarrassed, as with their lack of experience they would be unable to understand the motives for the crime and how it was committed. The geological reconstruction of the past appears like a police investigation: faced with a mountain chain, the geologist accumulates clues for the identification of the culprits and their motives. How did two plates become telescoped into each other? What are the deeper causes? Before the advent of plate tectonics, this exercise was as difficult as that of the policemen in an ideal world, as we did not know how to interpret the geodynamic events taking place before our very eyes. Since then, geodynamics has furnished us with a number of scenarios which may be applied with different degrees of success to situations of the past.

In this respect, the ophiolites present a serious problem. We do not know with certainty of any place on the globe where an ophiolite is actually emerging from an ocean. The models which we have outlined in Chapter 5, although solidly anchored on geological facts, do not possess the same reliability as, for instance, an example of crustal slivers scaling off along the East Pacific Rise would have. Why should this be so? Should we revert to the persistent doubt of assuming that the ophiolites are natural freaks, the study of which would contribute little to the understanding of the oceans, just like knowing about a unicorn would help little in understanding the anatomy of the horse.

We shall show in this chapter that the "production" of ophiolites in the oceans must take place during certain privileged moments of plate tectonics, which would imply that this is not in an absolutely continuous fashion, but either by evolving with time in an irreversible manner or in cycles. These concepts are by no means new and the existence of an old continental block, Pangea, which is now dispersed into six continents, is good evidence to some authors for an evolution in large terrestrial cycles referred to as Pangean.

A Supercontinent at the Dawn of the Mesozoic

The jigsaw puzzle of the six continents still combined some 200 Ma ago in a supercontinent, the Pangea (Fig. 9.1), was a concept already presented by Wegener at the start of the century. Initially impressed by the fit of the coast lines of the continental plates, Wegener built his arguments mainly on paleontological data. Thus the small fossil reptile Mesosaurus, which lived 270 Ma ago, was found only in Brazil and South Africa. From this he concluded that the two continents were continuous at some time prior to their present state of separation. The continuity of trends of ancient mountain ranges between the African and South American coastlines also puzzled geologists (Fig. 9.2). We know that this initially strongly rejected mobilistic view became firmly established with plate tectonics. We shall discuss three of the main arguments: the fit of the continents, paleomagnetism, and last but not least, plate kinematics.

The fit of the continents, which is already quite obvious when just looking at the map, is notably improved if, instead of trying to adjust only the coastlines, we take the true boundaries of the continents which lie along the edge of the continental shelf in the fairly steep

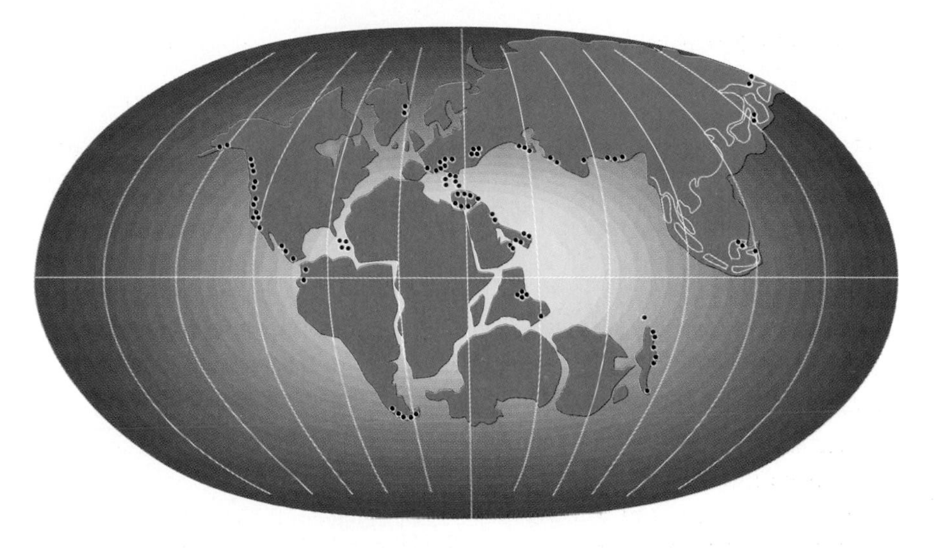

Figure 9.1
The shores of Pangea 200 Ma ago and position of the younger ophiolites. Note that except for the Caribbean and western Mediterranean ophiolites, these formations became obducted along the edge of the supercontinent. (After Abbate et al. 1985, Ofioliti, 10, 109–138)

topography of the continental slope. We know that when the opening of the oceans started to split the fragments of Pangea, like the Atlantic separating Eurasia and Africa from the Americas, the break or rift opened in the slope region. The continental shelf became submerged because of the stretching accompanying the break and of the subsequent subsidence (Chap. 3). The coastline itself is of little significance, as it tends to fluctuate due to events like the melting of ice. Compared to Wegener's reconstruction, the fit of Fig. 9.3 is improved by the use of a computer, which rigorously minimizes gaps and overlaps between the continents.

Paleomagnetism permits us to follow the drift of continents over geological periods. It actually records the terrestrial magnetic field in rocks at the time of their formation. Let us imagine a sedimentary rock forming: some of its mineral constituents sinking to the bottom are susceptible to the terrestrial magnetic field and align themselves to it like the needle in a compass. After its consolidation, the sediment may preserve a souvenir of this magnetization. It will form a fossil compass from which in the laboratory the direction of the poles may be reconstructed and to some extent the position of the rock on the globe at the time of its formation. The method also applies to vol-

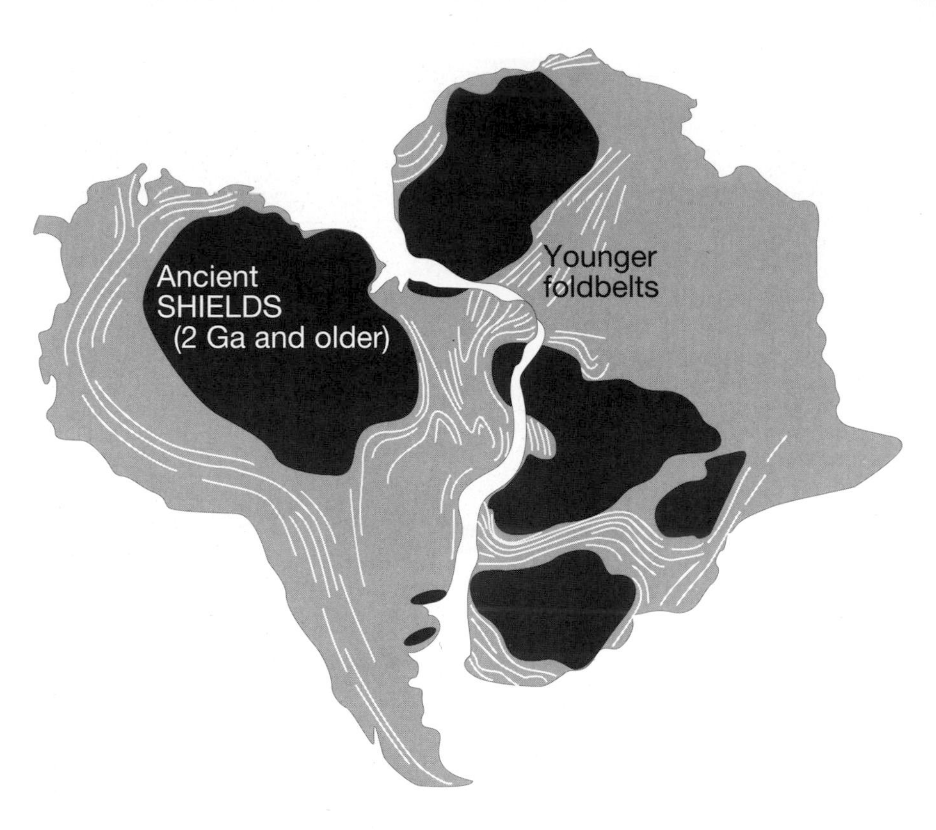

Figure 9.2
The extension of geological structures between Africa and South America, demonstrating the coherence of the two continents prior to the breakup of Pangea. The ancient shields, also referred to as cratons, are bordered by fold belts which are older than 450 Ma. (After P. Hurley 1979, La dérive des continents, Bélin Ed., 20–31)

Figure 9.3
When a computer solves a puzzle. The near-perfect fit of the borders of the continental plates (in *dark orange* zones of overlap, in *mauve* the gaps) is an excellent demonstration of the reality of Pangea. (M. E. Bullard 1979, La dérive des continents, Bélin Ed., 45–55) ▶

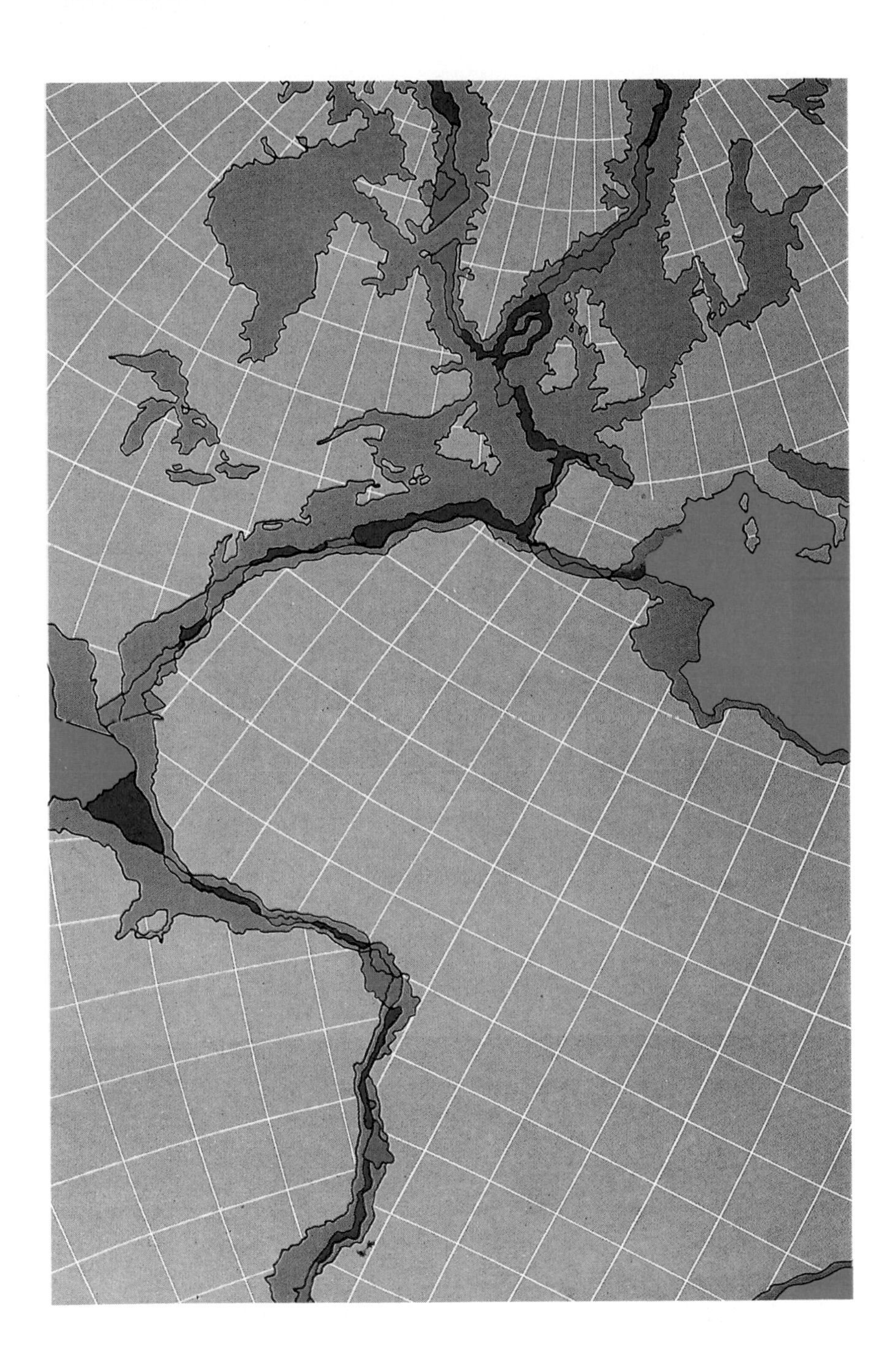

Figure 9.4
Trajectories of the magnetic poles from the continents constituting Pangea. The trends overlap from the Carboniferous (*C*) to the end of the Triassic (*T*) indicating that the continents were welded together into Pangea and drifted together. Note the divergence of the trajectories from the Triassic onwards. (After M. W. Mc Elhinny 1973, Palaeomagnetism and plate tectonics, Cambridge Univ. Press ed., 357 p.)

canic rocks which fix the magnetic field when cooling. By studying the paleomagnetism of selected sedimentary of volcanic formations, the age of which has been determined by fossils or radiometric dating, we may trace the drift of a continent. Paleomagnetism thus is a very valuable tool demonstrating the existence of Pangea and tracing its coalescence during the Carbonitferous (250–300 Ma) and then its dispersion from the Triassic onwards, 150/200 Ma ago (Fig. 9.4). As long as they drifted jointly as part of Pangea, the various continents exhibit the same paleomagnetic signature. The moment they started to diverge, the magnetic records also diverge just as they had been converging prior to the formation of Pangea.

The kinematics of oceanic plates are based on the study of the opening of oceans from magnetic anomalies. We have described the principle behind this in Chapter 1. Starting from the map of magnetic anomalies in which thc agc incrcascs as we move away from the respective ridge, we may close up the ocean progressively by removing one anomaly after the other in increasing age along the ridge. As in a film run backwards, the last scene on the screen will present the reconstruction of Pangea (Fig. 1.5).

And Tomorrow?

Let us return to our film of ocean kinematics in the proper sense: the North Atlantic opened after 165 Ma ago (Fig. 1.5) pushing Africa towards the east and from 90 Ma onwards northeast, whence it started to encounter Eurasia (Fig. 1.4). The two continents then became welded together locally along the Alpinc chain. We are able to make certain predictions about the future displacements of the plates which, as in meteorology, decrease in reliability as we move further into the future. In this context, we have already predicted the Iran-Oman collision and the crushing of the Oman ophiolites as happening some 2 Ma from now (Chapt. 5) with, as we believe, very little error margin. But who will be able to judge?

In the long term, the renewed closing of the Atlantic Ocean appears probable. The lithosphere along its North America and European margins has attained an age close to or above the critical limit of 150 Ma beyond which the oceanic lithosphere will be so thick and heavy that it could be subducted spontaneously (Chap. 1). We may thus predict the appearance of subduction zones in the geologically near future along those parts of the margins where the oldest lithosphere is found. The North Atlantic will then progressively close up at least as long as the Mid-Atlantic Ridge does not compensate the effect of these subductions by achieving a much higher spreading rate. This would be rather unlikely, as we know from seismic tomography that the mantle below this ridge is comparatively cool. Similar subduction zones should develop a few tens of million years later along the South American and African margins which started to open 120–130 Ma ago. With this closure, a new Pangea would take shape. Would this Pangea be the last one, or are there reasons to believe that it would also start to disperse at a later date?

The Pangean Cycle

An answer to this question is furnished by thermal consideration. We know that under the continents the lithosphere is particularly thick. It constitutes a "thermal shield" through which the heat coming from the deep mantle can escape only with difficulty (Fig. 9.5). This part of the mantle will thus heat up progressively, until below the vast cover of Pangea an "explosive" situation is established over a time span of 100 Ma. The Pangea becomes uplifted by the hot mantle, is stretched in the process, and starts to crack along linear rifts, some of which might follow lines of weakness representing scars between ancient continents which are now welded together. Cut up by these rifts, the new continents slide into the Panthalassa Ocean (combined sea), starting a new phase of dispersion.

Dispersion and agglomeration of continents would constitute the two main phases of a Pangean cycle. With the present, considerable dispersion of the continents, we find ourselves within the middle of a cycle which, with the period of dispersion covering about 150 Ma, will be followed by a phase of regrouping extending over the same time span. As the reconstructed Pangea has a life span in the order of 100 Ma, as we shall see, the complete cycle will encompass some 400 Ma.

There are paleomagnetic and geological arguments in favour of this concept of major 400 Ma cycles in the history of the Earth, but this enticing concept cannot be considered as proven. In paleomagnetic circles, a Paleozoic cycle is discussed during which a supercontinen was to have started its dispersion between 600–500 Ma ago, i. e. 300–400 Ma before the start of the dispersion of our Pangea during the Jurassic. The study of mountain ranges, like the Alps, sometimes revealed moreover that they represent scars left by repeated openings and closures. A certain periodicity with a "wavelength" of sever-

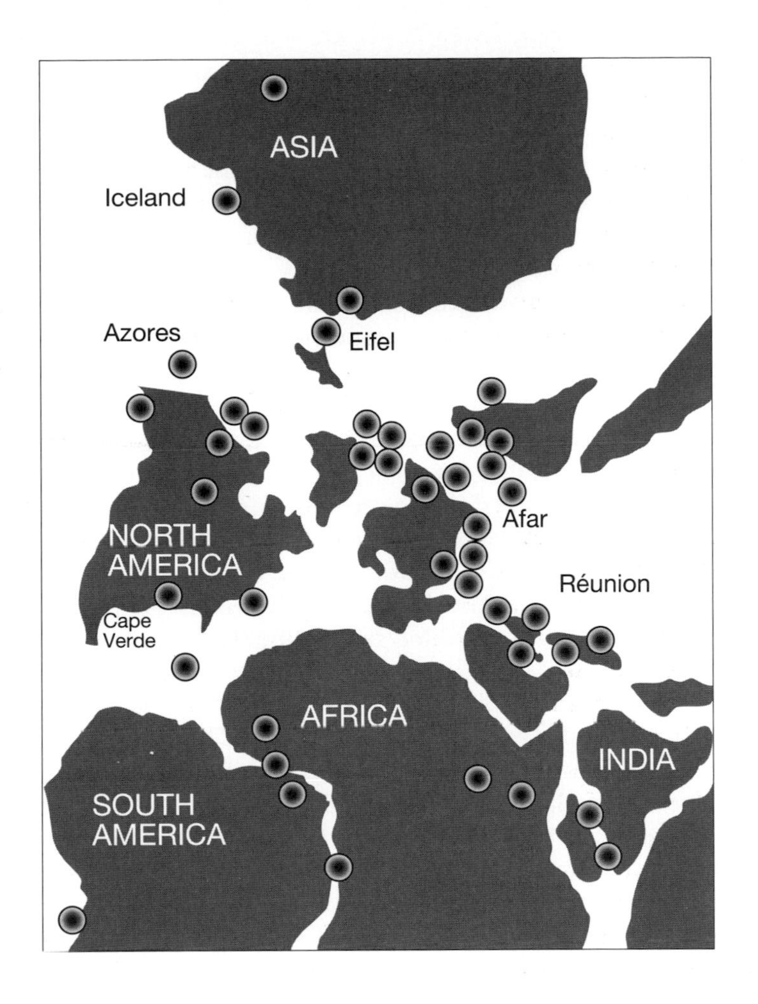

Figure 9.5
The Pangean thermal shield and the origin of certain hotspots, after the American geophysicist D. Anderson. The hotspots, located by *circles* in their present position on the planispheric projection, would have been generated by the accumulation of heat below the Pangean thermal shield. Actually, when the map of the hotspots is superimposed onto that of Pangea 300 Ma ago, we find that the latter covers most of the hotspots and that their concentration is higher where the centre of Pangea was at the time. (After D. L. Anderson 1980, Episodes, 3, 3–7)

al hundred million years is also suggested by changes in sea level and in rock chemistry which may be related to a cyclical activity of ridges. Strong activity signifies the vast creation of new lithosphere at oceanic ridges with a pronounced relief and thereby the flooding of the oceans expressed by marine transgression onto the margins of the continents and into their interior basins. Such a vast transgression took place during the Cretaceous 100 Ma ago. Inversely, weak ridge activity will be felt by a regression of the sea level. We would like to draw attention to the close cooperation here between geodynamics, the youngest branch of earth sciences, and its oldest member, namely stratigraphy, which unravels the sedimentary records accumulating on the continental margins and in the intracontinental basins.

Another possible cause for cyclic events in the Earth history would be the "avalances" toward the lower mantle of large masses of lithospheric slabs accumulated above the 670-km discontinuity (Fig. 2.5). Such huge and sudden events would induce a catastrophic turnover of mantle convection with equally large volumes of deep and hot mantle surging up (the so-called superplumes).

Ophiolites Created at Springtime

Age determinations on more than 200 ophiolites reveal trend suggesting the "production" of ophiolites at certain specific points in time (Fig. 9.6). Data for a first period rally around 500–400 Ma and a second better defined period from 150–50 Ma. The peaks in Fig. 9.6 are somewhat widened and the older one rather "eroded". This is explained by the difficulties encountered in dating older rocks, the age of formation of which has frequently been altered by later metamorphic events.

Within the framework of the hypothetical Pangean cycles, the formation of ophiolites takes place during the early phase of such a cycle, as clearly obvious in the youngest cycle. The diagram also shows that the present phase of wide continental dispersion is by no means conducive to the formation of ophiolites. Why would ophiolites form during the spring of a Pangean year?

In Chapter 5, where we have described how ophiolites come to lie on the continental margins, we have pointed out that the oceanic lithosphere represented by these ophiolites is usually rather young. Let us recall that in contrast with old lithosphere, which tends to become subducted because of its thickness and weight, the very young and light lithosphere cannot be subducted. During the convergence of plates it can override the continental margins and come to lie on them in the form of ophiolites. This process is called obduction. The dispersion of Pangean starts by the opening of new oceans. We have pointed out above that this expansion of the ocean culminated during the Mid-Cretaceous. The ophiolites would be stranded ashore preferentially on the "high tide of the lithosphere", itself responsible for the rise of the waters onto the continents and the vast transgression experienced during this period.

When the ridges are particularly active, the subduction zones should be equally so, as otherwise the surface of the Earth and consequently also its radius should increase. This is, however, clearly not the case. Young and active ridges combined with equally active subduction zones represent a favourable environment for the detachment of young lithosphere and its obduction. On the map of Pangea (Fig. 9.1), we have pointed out ophiolites emplaced during the dispersive phase. We observe now that ophiolites occur around the circumference of Pangea. They actually outline the trace of subduction zones, as in this configuration the rifts and the future ridges are to develop within Pangea, whereas the subduction zones spread around the outer edge.

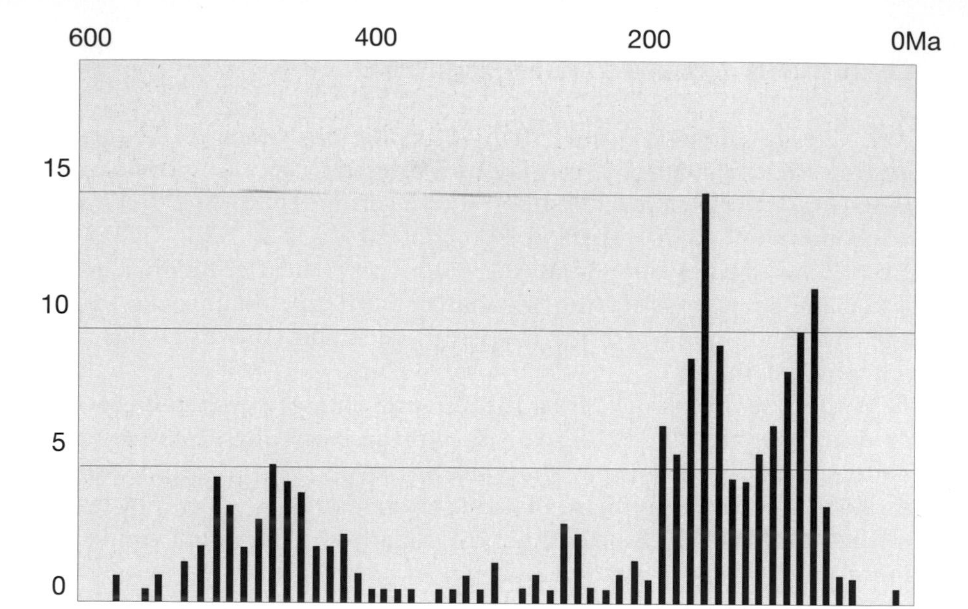

Figure 9.6
Age frequency of ophiolites. Ophiolites formed mostly around 500 and 100 Ma ago. The peak around 150 Ma may be ascribed to the over-representation of age determinations on Jurassic ophiolites of the western Mediterranean. (After Abbate et al.1985, Ofioliti, 16, 109–138)

Since plate tectonics, as we know it now, has kept the surface of our planet in motion for at least 2000 Ma, we are justified in believing that the Earth lives in slow pulsations of about 400 Ma duration, during which successive Pangeas come and go. The ophiolites, then, would be stranded on the shores of a Pangea during its dispersive phase.

Glossary

accretion: an addition of material. Along zones of subduction, e. g. the active margins of continents, continental crust may accrete. In cases of expansion, one sometimes refers to rifts and ridges as zones of accretion, in this case of oceanic crust. It is also said that under the ridges the lithosphere accretes through cooling of the asthenosphere.

adiabatic (system): an adiabatic system does not exchange heat with its surrounding environment. Thus, a convecting asthenosphere is ideally an adiabatic system. This system is not isothermal, it possesses an adiabatic gradient as with increasing depth and pressure its temperature should also increase.

amphibole: silicate with ribbon structure obtained by the coalescence of two simple chains of the pyroxene structure, allowing the entry of water into the lattice. Chemically, the amphiboles are rather close to the pyroxenes from which they can be derived by hydration.

amphibolite: metamorphic rock developing from basalt at recrystallization temperatures of about 450–750 °C. In a more general way, we talk about amphibolite facies when dealing with rocks of whatever chemical composition which resulted from metamorphism within this temperature range.

andesite: volcanic rock containing more silica and alkaline elements (sodium, potassium) than basalt. Although rare in the oceanic crust, it is the typical rock of the island arcs.

asthenosphere: part of the upper mantle situated below the lithosphere. It is sufficiently hot (1100 °C or above) to be soft (plastic) and is pervaded by very slow convection currents.

back-arc basin: an oceanic basin lying beyond an island arc, separated by it from the oceanic trench, and consequently located above the deeper zones of the subduction zone. The oceanic expansion of these basins is one of the secondary effects of subduction.

basalt: the main volcanic rock, resulting from the partial melting of mantle peridotites. Its black colour is caused by its relatively high content of iron and titanium when compared to other volcanic rocks. Basalt consists of plagioclase, pyroxene and frequently also olivine. It contains less silica (about 50 %) than most other volcanic rocks.

boundary layer: relatively narrow inert layer between two convection cells. Heat is transferred here by conduction.

chromite: oxide of the spinel family rich in chrome. The chromite of ophiolites is alway associated with the dunites. Disseminated in peridotite, it is derived from very great melting of basalt as a refractory residue alongside olivine. Concentrated in chromite deposits, it may result from an early crystallization of basalt.

clinopyroxene: → pyroxene.

conduction: transport, in particular of heat, by diffusion within a certain environment. Because of the slow rate of this type of propagation, conduction in rocks represents a mode of heat transfer of low efficiency.

conductive (gradient): → geothermal (gradient).

convection: slow movement of thermal origin in the mantle with transfer of heat by moving material. Considering the transport velocities in the mantle, this transfer process is much more efficient than conduction.

convection currents: vast movement of material in the mantle in the form of convection cells, the dimensions of which range up to several thousands of kilometres and the velocities, to a few centimeters per year.

cumulate: generally stratified magmatic rock, the origin of which is attributed to a sedimentation (accumulation) of crystals within a magma chamber.

diapir: ascent of lighter plastic rocks within heavier formations, like fume from a stack, i. e. warmer and lighter air rising in calm, cool air. Mantle diapirism under the ridges will result from melting which lowers the density and increases the fluidity of the material affected.

diapirism: → diapir.

diopside: → pyroxene.

dunite: peridotite made up of olivine and spinel (chromite). Ultimate residue of partial melting when associated with harzburgites and lherzolites. Possibly locally magmatic in origin on the floor of magma chambers.

dyke complex: system of dykes of basaltic composition aligned and parallel to each other. They intrude upon each other under the axes of ridges and rifts.

dynamics: description of the effects of forces, opposed to kinematics which deals with motion
elements (major): → geochemistry.
elements (minor): → geochemistry.
enstatite: → pyroxene.
fault (listric): normal fault, i. e. a fault in which the section above the usually inclined fault plane slides downwards, and the inclination of which becomes increasingly flatter with depth. It leads to a corresponding rotation of the overlying portion (→ Fig. 3.15).
fault (normal): fault in which the portion above the fault plane (if this is inclined) slides downwards (→ Fig. 3.13). In the case of a *reverse* fault, it move upwards. A normal fault contributes to horizontal expansion, the more so when its inclination becomes flatter.
fault (transform): major fault dissecting the lithosphere and facilitating the contact between two plate boundaries. Between, e. g. a ridge and a subduction zone, it "transforms" expansion into compression. It is vertical and the movement along it is horizontal.
fault (transverse): a fault which is arranged transversally across a ridge, dissecting it and interrupting its continuity.
feldspar: light silicate, rich in silica and alumina. We distinguish the plagioclase series containing notably calcium and sodium from the alkaline feldspar series rich in potassium and sodium. They represent the most abundant minerals of the crust. Their light colour is due to the absence of iron and titanium.
flux (heat): quantity of heat travelling through a unit surface area per second. It is usually very low, except in the vicinity of volcanoes.
foliation: plane of mineral flattening in deformed rocks.
fracture zone: describes the large submarine transverse or transform faults. They derived their name from their peculiar topographic expression and the deformed rocks dredged from them.
gabbro: rock resulting from the slow crystallization of a basalt at depth when the constituting minerals (plagioclase, pyroxenes and olivine) have time to grow to sizes visible to the naked eye.
garnet: dense silicate of dark colour due to relatively high iron content, containing also magnesium, calcium, alumina, and chrome. It is common in metamorphic rocks and is also found in peridotites equilibrated at depths greater than 75 km.
geochemistry: subject dealing with the chemistry of rocks. We distinguish between the geochemistry of the major elements Si, Al, Fe, Mg, Ca, Na, K, Ti and that of the trace elements which generally do not exceed 1 % of the rocks, and isotope geochemistry.

Measuring the proportions of the stable and the radiogenic isotopes of certain elements permits us to trace in detail the origin and the evolution of these rocks and to measure their absolute ages.

geochemistry (isotopic): → geochemistry.

geochronological dating: → geochronology.

geochronology: group of methods employed to date rocks. Relative geochronology relies on fossils, whereas absolute geochronology makes use of the radioactive decay of certain isotopes (→ also geochemistry).

geochronology (isotopic): → geochronology.

geodynamics: field of the earth sciences dealing with the movements and forces shaping the globe.

geoid: surface corresponding to the mean ocean level. Anomalies of the geoid are expressed by differences in elevation between the geoid and a reference ellipsoid.

geotherm: → geothermal.

geothermal (gradient): variation in temperature with depth within the Earth. The corresponding curve is called a geotherm. Within stationary domains the gradient is conductive (→ conduction), whereas in moving domains it is convective and generally adiabatic (→ convection).

gravimetry: field of geophysics dealing with the distribution of masses at depth (gravity) on the basis of variations in the gravitational force of the Earth as measured on the surface by a gravimeter or in space with satellite technologies.

gravity: → gravimetry.

greenschist: metamorphic rock frequently formed from basalt or gabbro at temperatures between 150–450 °C. At still higher temperatures the greenschists recrystallize to amphibolites. The term greenschist facies describes the group of rocks subjected to metamorphism at this level.

harzburgite: peridotite with olivine, orthopyroxene and spinel resulting from a large extraction of basaltic liquid.

hotspot: oceanic island or continental region characterized by abundant volcanism and exceptionally high relief and thermal flux.

island arc: a system of volcanoes arranged above a subduction zone. The ring of fire around the Pacific is made up of such arcs.

isotopic composition: → geochemistry.

kinematics: analysis of motion. Plate kinematics: relative movement of plates. Kinematic analysis: study of the movement of rocks on the basis of their deformation structures.

lherzolite: peridotite with olivine, orthopyroxene, clinopyroxene and either garnet (equilibration beyond 75 km depth), spinel (equilibration between 30–75 km) or plagioclase when equilibrated above 30 km. Lherzolite represents the normal upper mantle facies prior to the extraction of basalt.

lineation: trace observed on the foliation plane of deformed rocks. Alignment of minerals in this plane defines a mineral lineation corresponding in general to the direction of stretching.

lithosphere: crust and cold, rigid parts of the mantle. The plates consist of lithosphere. Its thickness increases with age up to about 100 km in the oceanic domain.

magma chamber: space filled with magma. Under ridges and rifts, such chambers may develop on a permanent or temporary basis. Slow and rhythmic crystallization of basalt here usually leads to banded gabbros.

magnetic anomaly: the difference between the measured and the expected magnetic field of a given area in the absence of local disturbing factors. The magnetic anomalies of the oceans define alternating positive bands caused by rocks with normal magnetization, i. e. parallel to the present field, and negative bands resulting from rocks with inverse magnetization.

magnetism: field of geophysics studying the magnetic properties of rocks (magnetization) and natural systems (magnetic field, etc.).

mantle: shell of the earth limited by the base of the crust at a depth of about 10–30 km and by the core at a depth of 2900 km. Below a depth of 400 km the peridotites, which make up the mantle, start to transform into denser rocks under the influence of increasing pressure.

margin (continental): border between continental and oceanic crust, not to be confused with the shoreline which generally runs further in the direction of the continent's interior. The margin is referred to as active where it is marked by a subduction zone like, e. g. the Andean margin, or passive, e. g. the European-Atlantic margin. Its border then coincides with the continental slope.

metamorphism: group of transformations taking place in rocks due to an increase in temperature alone or in combination with a concurrent increase in pressure (burial). Metamorphism is frequently accompanied by deformations.

Moho: boundary between crust and mantle at a mean depth of 6 km below the oceans and about 30 km below the continents.

obduction: transport of a fragment of oceanic lithosphere (ophiolite) onto a continental margin.

olivine: silicate of magnesium and iron – $(Mg_{0.9}, Fe_{0.1})_2SiO_4$ – the main constituent of upper mantle peridotites.

ophiolite: fragment of oceanic lithosphere (crust and mantle) emplaced on the continent and frequently part of mountain belts.

orthopyroxene: → pyroxene.

overlapping spreading centre (OSC): border zone between two ridge segments, the ends of which are drawn out and turned against each other (→ Fig. 3.1). They are observed especially along the East Pacific Rise.

paleomagnetism: study of the terrestrial magnetic field as recorded by the magnetization of rocks during their formation. Having determined the fossil field, one can reconstruct the displacements and rotations suffered by this rock since its origin.

peridotite: this is the rock of the upper mantle down to a depth of 400 km. Its composition is presented in the table on p. 106. It contains predominantly olivine, either together with spinel alone (-dunite) additionally with orthopyroxene (harzburgite) or with orthopyroxene and clinopyroxene together (lherzolite). In the latter case, the accessory presence of garnet indicates a provenance from a depth below 75 km, that of spinel a provenance from 30–75 km, and that of plagioclase a provenance shallower than 30 km.

plagioclase: → feldspar.

plagiogranite: oceanic granite depleted in potassic minerals (black mica, alkali feldspars) against continental granites.

plate: entity of lithosphere partly covering the surface of the Earth and delineated by rifts or ridges, subduction zones and/or transform faults.

plume: jet of hot material ascending through the mantle and leading to volcanic hotspots on the surface.

pyroxene: silicate with a chain structure. Orthopyroxene or enstatite contains virtually only iron and magnesium – $(Mg_{0.9}\ Fe_{0.1})_2\ Si_2O_6$. Clinopyroxene or diopside contains additionally calcium, alumina, and sodium – $(Mg_{0.9}\ Fe_{0.1})$ (Ca, Na) $(Si, Al)_2O_6$. Melting of the diopside in fertile mantle peridotites largely contributes to the formation of basalts.

ridge (oceanic, mid-oceanic): submarine mountain range along which oceanic plates are formed. Called mid-oceanic when the ridge opens as a continuation of a rift originating within a continent.

rift (oceanic, continental): elongate depression between relatively raised shoulders marked by volcanism and tectonic stretching. The rifting structure is symptomatic for relatively slow expansion.

The rift may be continental (Upper Rhine Graben) or oceanic (Atlantic ridge).

segmentation: In the case of oceanic ridges, this is the longitudinal subdivision of the ridge into segments by transform faults or overlapping spreading centres.

segments: → segmentation.

seismics: → seismology.

seismology: field of geophysics dealing with the propagation of seismic waves through the globe. The waves are induced by earthquakes. They may be used for unravelling the internal structure of the Earth.

serpentine: family of minerals forming from olivines and pyroxenes through the addition of water, also called hydrothermal alteration at temperatures from around 400–800 °C (ridge hydrothermalism) to ambient temperatures (later or recent oceanic alteration). The serpentinites, rocks rich in serpentine, are derived from peridotites in a process referred to as serpentinization.

serpentinite: → serpentine.

serpentinization: → serpentine.

shelf (continental): an area along a passive continental margin, the depth of which does not exceed a few hundreds of metres, extending between the shoreline and the slope.

slope (continental): boundary between continent and ocean along passive margins. Situated at the edge of the continental shelf, the slope is notably due to its relatively steep gradient. Its base is defined by the vast abyssal plains several thousands of metres deep.

spinel: oxide made up of chromium, iron, magnesium and alumina present in lherzolites to a depth of 75 km. Not to be confused with the transformation into the spinel phase by olivine beyond a depth of 400 km.

subduction: plunging of oceanic plates along subduction zones localized either within the oceans or along their boundary with the continents. Plates descend into the mantle at different angles down to at least 700 km.

tectonics: field dealing with terrestrial deformations on any scale, here in particular deformation of the plates.

tomography (seismic): fairly new technique that delineates in the interior of the Earth domains in which the seismic waves exhibit abnormal velocities which may be associated with variations in temperature, composition or structure. A highly complicated technique availing itself of the same principles as medical tomography.

wehrlite: Magmatic rock rich in olivine, clinopyroxene and sometimes plagioclase. It forms intrusions in the crustal part of ophiolites.

Source of Illustrations

Chapter 1
E. Ball: 1.2–1.4, 1.6–1.12

Chapter 2
P. Thomas: 2.1
J. P. Montagner: 2.2
E. Ball: 2.3, 2.4, 2.6–2.9
R. I. Tilling et al.: 2.10, 2.11

Chapter 3
K. C. Macdonald et al.: 3.1, 3.4
E. Ball: 3.2, 3.5, 3.6, 3.8, 3.10, 3.11, 3.13
A. Cazenave: 3.3
J. M. Auzende: 3.7
W. P. Irwin and R. G. Coleman: 3.9
P. Gente: 3.12
C. Burchfield: 3.14
J. L.Faure and J. C. Chermette: 3.15

Chapter 4
E. Ball: 4.1, 4.5
J. M. Auzende: 4.2–4.4

Chapter 5
A. Nicolas: 5.1, 5.6, 5.7
E. Ball: 5.2–5.5, 5.7g, 5.8

Chapter 6
A. Nicolas: 6.1, 6.3, 6.4, 6.8
E. Ball: 6.2, 6.5–6.7, 6.9–6.11
M. P. Ryan: 6.12

Chapter 7

E. Ball: 7.1–7.3, 7.5–7.9, 7.11, 7.12
C. Talbot and M. Jackson: 7.4
F. Boudier: 7.10

Chapter 8

E. Ball: 8.1, 8.3–8.7, 8.9
G. Féraud: 8.2

Chapter 9

E. Ball: 9.1, 9.2, 9.4–9.6
E. Bullard: 9.3

In Further Collaboration

E. Ball, A. Cinçon, S. Fournier, J. Goyallon, D. Pauvert, C. Repérant

Springer-Verlag and the Environment

We at Springer-Verlag firmly believe that an international science publisher has a special obligation to the environment, and our corporate policies consistently reflect this conviction.

We also expect our business partners – paper mills, printers, packaging manufacturers, etc. – to commit themselves to using environmentally friendly materials and production processes.

The paper in this book is made from low- or no-chlorine pulp and is acid free, in conformance with international standards for paper permanency.